£2.25

ELEMENTARY ELECTROMAGNETIC THEORY

VOLUME 1
STEADY ELECTRIC FIELDS AND CURRENTS

BOOKS BY DR. CHIRGWIN, DR. PLUMPTON AND
PROFESSOR KILMISTER ON ELECTROMAGNETIC THEORY

ELEMENTARY ELECTROMAGNETIC THEORY

Volume 1   *Steady Electric Fields and Currents*
Volume 2   *Magnetic Fields, Special Relativity and Potential Theory*
Volume 3   *Maxwell's Equations and Their Consequences*

ELEMENTARY ELECTROMAGNETIC THEORY

VOLUME 1
STEADY ELECTRIC FIELDS AND CURRENTS

B. H. CHIRGWIN
Queen Mary College, London

C. PLUMPTON
Queen Mary College, London

AND

C. W. KILMISTER
King's College, London

PERGAMON PRESS
OXFORD · NEW YORK · TORONTO
SYDNEY · BRAUNSCHWEIG

Pergamon Press Ltd., Headington Hill Hall, Oxford
Pergamon Press Inc., Maxwell House, Fairview Park, Elmsford, New York 10523
Pergamon of Canada Ltd., 207 Queen's Quay West, Toronto 1
Pergamon Press (Aust.) Pty. Ltd., 19a Boundary Street,
Rushcutters Bay, N.S.W. 2011, Australia
Vieweg & Sohn GmbH, Burgplatz 1, Braunschweig

Copyright © 1971 B. H. Chirgwin, C. Plumpton, C. W. Kilmister

All Rights Reserved. No part of this publication may be reproduced, stored in a retrieval system, or transmitted, in any form or by any means, electronic, mechanical, photocopying, recording or otherwise, without the prior permission of Pergamon Press Ltd.

First edition 1971

Library of Congress Catalog Card No. 70-129631

London Borough of Enfield Public Libraries

J34906

530.141

Fo 64847

Printed in Hungary

08 016079 4 Hard cover
08 016080 8 Flexicover

CONTENTS

Preface to Volume 1   vii

1. Introductory Concepts   1

    1.1 Preliminary physical ideas   1
    1.2 Method and aim   17
    1.3 Scheme of development   21
    1.4 Systems of units   22

2. The Inverse Square Law in Electrostatics   23

    2.1 Charge   23
    2.2 The electric field and electric potential   24
    2.3 Electric displacement   26
    2.4 Coulomb's inverse square law   29
    2.5 Two-dimensional fields   36
    2.6 The field at internal points   38
    2.7 Gauss's theorem   40
    2.8 Discontinuities and boundary conditions   51
    2.9 Applications of Green's theorem   55
    2.10 Dipoles and multipoles   58
    2.11 Double layers   61
    2.12 Volume distributions of dipoles: polarization   64

3. Conductors in the Electrostatic Field   71

    3.1 The property of a conductor   71
    3.2 Some general theorems   72
    3.3 Systems of conductors: capacitance   79
    3.4 Forces and energy   90
    3.5 Criticisms of the expression for energy   103
    3.6 The location of electrostatic energy   105

4. Dielectrics   111

    4.1 The effects of a dielectric   111
    4.2 The general theory of dielectrics   114
    4.3 The uniqueness and reciprocal theorems   120

4.4 The energy of the electrostatic field	130
4.5 Minimum energy theorems	137
4.6 The electrostatic stress system: Maxwell's stress tensor	141

5. The Steady Flow of Electric Currents — 149

5.1 Introduction	149
5.2 The equation of continuity (conservation) of charge	150
5.3 Ohm's law	151
5.4 Boundary conditions and discontinuities	153
5.5 The rate of heat production	155
5.6 Electromotive force	156
5.7 The analogy with the electrostatic field	158
5.8 Integral theorems	166
5.9 Networks of linear conductors	172
5.10 Integral theorems applied to networks	183

ANSWERS TO THE EXERCISES — 191

INDEX — 195

PREFACE TO VOLUME 1

THIS is the first of three volumes intended to cover the electromagnetism and potential theory usually included in an undergraduate's course of study. These books are intended only as an introduction to electromagnetism and have been prompted by discussions with first, second and third year undergraduates.

In this first volume, we consider the experimental results which require mathematical explanation and discussion, in particular those referring to phenomena which suggest that the simple Newtonian concepts of space and time are not fully valid. Then we consider steady-state fields and deal next with electrostatics, including dielectrics, energy theorems, uniqueness theorems, and end this volume with a chapter on the steady flow of electric currents. Throughout SI units are used although the older systems are briefly mentioned.

The general scheme of the volumes is to start with the simple case of steady fields and to develop the appropriate generalizations when this constraint is relaxed. Thus, in the first volume we start with the fields associated with stationary charges and relax the stationary condition to allow consideration of the flow of steady currents in closed circuits.

In the second volume we first consider the magnetic field of steady currents—magnetostatics. Relaxing the constraint of stationary steady currents we are led to consider electromagnetic induction when the current strengths in closed circuits vary or when the circuits move. This leads to the necessity of considering the breakdown of Newtonian ideas and the introduction of special relativity. When we further relax the constraint of closed circuits and consider the motion of charges in open circuits we are led to introduce displacement current because of the relativistic theory already set up, and so are led to Maxwell's equations.

In the third volume, we consider the implications of Maxwell's equations such as electromagnetic radiation in simple cases, and deal further with the relation between Maxwell's equations and the Lorentz transformation.

We assume that the readers are conversant with the basic ideas of vector analysis including vector integral theorems.

Our thanks are due to the University of Oxford, to the Syndics of the Cambridge University Press and to the Senate of the University of London for permission to use questions set in their various examinations.

B. H. CHIRGWIN
C. PLUMPTON
C. W. KILMISTER

CHAPTER 1

INTRODUCTORY CONCEPTS

1.1 Preliminary physical ideas

Before embarking upon an exposition of theoretical electromagnetism we state briefly the physical ideas and concepts with which the reader is assumed to be familiar. Roughly, these ideas are covered in school physics courses. Further, we assume that the reader is familiar with a number of ideas and methods of mathematics and mechanics; chief among these for our present purpose is an understanding of vectors, vector algebra (see *C.M.*,[†] vols. 2 and 3) and vector analysis (see *C.M.*, vol. 4).

At present there is considerable discussion among pure scientists and technologists concerning the system of units most suitable for use in electromagnetic theory. Since the Système International (SI) appears likely to become generally adopted, especially by technologists, we use the SI in this volume. (However, in this chapter, where we refer to experiments and measurements originally made before the systematic use of SI, we must refer to the units used in these experiments, viz. CGS units of the electrostatic and electromagnetic system.)

We give below under five headings a summary of the ideas and topics assumed as preliminary knowledge on the part of the reader.

1. MECHANICS

These ideas comprise Newton's laws of motion and the concepts of force, position, velocity, acceleration, expressed with the aid of vectors. The units in which these quantities are measured are based on the units of length, mass and time which are arbitrarily chosen to be as shown in Table 1. Other mechanical units are defined in terms of these basic units.

[†] B. H. Chirgwin and C. Plumpton, *Course of Mathematics for Scientists and Engineers*, 2nd revised edn., 1970.

TABLE 1

Quantity	SI	
	Name	Abbreviation
Length	metre	m
Mass	kilogram	kg
Time	second	s

Velocity and acceleration are time-derivatives; force is defined by the acceleration it imparts, viz. $P = mf$. Such units and their names are shown in Table 2.

TABLE 2

Quantity	SI	
	Name	Abbreviation
Velocity	metre per second	m/s or ms^{-1}
Acceleration	metre per second per second	m/s^2 or ms^{-2}
Force	newton	N
Work, energy	joule	J
Power, rate of working	watt (= 1 joule per second)	W

In this volume the phrase "unit length" will imply one metre, and "unit mass" will imply one kilogram, and so energies will be expressed in joules, power in watts, etc.

2. Electrostatics

We assume that the reader is familiar with the elementary qualitative aspects of electrostatics and the simple experiments used to demonstrate electrostatic phenomena. Certain bodies can be charged by rubbing in a suitable manner and there are two kinds of charge called positive and negative. Bodies in the neighbourhood of another charged body experience forces, bodies carrying charges of like kind repel one another and of unlike kinds attract one another. The presence of a charge can be detected by an instrument whose operation depends upon the repulsion of like charges, viz. an electroscope. Qualitatively, bodies can be divided roughly into two classes, conductors and insulators; a charge is able to move on a conductor but is unable to move on an insulator. This ability of charges to move is illustrated by the process known as *influence* (or sometimes *induction*) in which the proximity of a charged body to an uncharged conductor causes positive and negative charges to be separated to different parts of the conductor. Although charges are free to move on a conductor they are not free to leave the conductor except in special circumstances causing sparks or other kinds of discharge.

There are certain (crudely) qualitative results which can be shown in elementary electrostatics, e.g. charge tends to be concentrated on those parts of a conductor which have the greatest curvature. One quantitative experiment concerns influence, namely Faraday's "ice-pail" experiment, in which a charged conductor inside a closed (or approximately closed) conductor separates like charges onto the outside and unlike charges onto the inside surface of the hollow conductor (the "ice-pail"). The experiment shows that no force acts on the original charged conductor and that contact between this and the inside of the "ice-pail" leaves the original conductor completely uncharged, all its charge having been transferred to the (outside of the) "ice-pail". This, in fact, is the only way of transferring the whole of the charge on one conductor to another conductor. It also shows that there is no "electric field" inside a hollow, closed conductor.

The arrangement of two conductors close together (one of them often being earthed) to form a capacitor, or condenser, e.g. a Leyden jar, parallel plate condenser, relies on the process of influence to hold the charge on one plate by means of the attraction of the equal and opposite charge appearing on the neighbouring plate. This device enables a larger charge to be placed on the first plate than would otherwise be possible (without a discharge occurring by a spark) in the absence of the second plate. The capacity of such an arrangement can be increased by inserting an insulator (called a *dielectric*) other than air between the conducting plates.

Previously (i.e. before the systematic use of SI), quantitative results in elementary electrostatics were based upon the definition of unit charge, viz.,

> "Two identical charges at unit distance (1 cm) apart in a vacuum which repel each other with unit force (1 dyn) are unit charges."

This is the CGS definition of the electrostatic (ES) unit of charge. (See § 2.1.) Such a definition is not used in the SI system. Having defined a unit of charge, the strength of the electrostatic field *E* at a point is defined as the force acting on a unit charge at that point. Because of this force exerted on a charge, work must be done when any additional charge is added to a body already charged, the higher the original charge the greater is the work required for the additional charges. This is the basis of the idea of potential which is defined thus:

The potential at a point is the work that must be done against the force of the field in bringing a unit charge to that point from infinity (or from some other standard position).

It is the potential of a conductor which determines whether or not it will discharge itself by causing a spark. The capacitance of a condenser (capacitor) is measured by the charge which, when put on one of the plates, raises its potential by one unit. The action of the capacitor is explained by the presence of an equal and opposite charge attracted to the other plate which tends to reduce the potential of the first plate.

3. Electric currents and electrolysis

This is probably the section of elementary physics which is most familiar to the reader and which he has studied more in its quantitative aspects than cother sections of electromagnetism.

An electric current in a conductor is the continuous passage of electric charge along that conductor. Such a current has two chief effects; it produces heat in the conductor and it affects a compass needle. In addition this conductor attracts or repels a second conductor which is also carrying a current. The quantitative information on these phenomena centres around Ohm's law, and the starting point of the definitions of the quantities involved is the unit of current. This is defined in the CGS system in terms of the magnetic effect it produces. To maintain a current in an electrical circuit requires the expenditure of energy because of the production of heat. This energy is obtained either from a battery which converts chemical energy, or from a generator which converts mechanical energy into electrical energy, the energy reappearing as heat in the circuit. The concept of potential difference, or voltage, relies on this expenditure of energy for its definition: the potential difference between two points is defined as the rate of expenditure of energy necessary to maintain unit current flowing between these two points. Remembering that an electric current consists of charges in motion, we see that potential difference so defined is the same as potential in electrostatics. The latter is the work required to move unit charge, whereas the former is the rate of working required to maintain unit rate of passage of charge. The third quantity which occurs in Ohm's law, resistance, distinguishes between conductors of different metals, a distinction unnecessary in electrostatics which is only concerned with the situation when the charges have moved into their equilibrium positions. In any system of units Ohm's law takes the form $V = RI$; as commonly used and in the SI, V is measured in volts, I in amperes, and then R is measured in ohms. (The resistance R, in general, varies with temperature for any conductor.)

A familiar source of electric power in elementary physics is the galvanic cell or lead accumulator. In these batteries substances go into solution in the form of ions which are charged atoms or groups of atoms free to move in the liquid. This process of solution involves the release or consumption of chemical energy and the ions, being charged, constitute an electric current when they move. The passage of a current through a solution is accompanied by the deposition or solution of substances at the electrodes. The chemical energy released in a galvanic cell is the source of electromotive power which drives the current against the resistance of the conductors in the external circuit and against the resistance to the motion of the ions in the solution. The latter is the internal resistance of the cell.

The chief significance of the laws of electrolysis is that they indicate that the charge carried by an ion is a small integral multiple of a basic charge; the multiple is equal to the valency of the ion and the basic charge is numerically equal to that on an electron. The total charge which must be passed through a solution in order to liberate one gram-molecule of a monovalent ion (one mole) when divided by Avogadro's number (the number of molecules in one mole) gives an estimate of this charge. In fact, Faraday's laws of electrolysis gave one of the earliest indications of the atomic nature of electricity, and of the magnitude of the electronic charge.

4. Magnetism

Historically, the theory of magnetism was developed from investigations of the interactions of permanent magnets and the earth's magnetic field. A permanent magnet was regarded as a dipole; the theoretically perfect magnet, unrealizable in practice, consists of two point-magnetic poles at the ends of a rigid body. The poles, of two kinds, N (or north-seeking) and S (or south-seeking), always occur in pairs, one of each kind on any body, and like poles repel, unlike poles attract one another. But it is impossible to obtain, certainly in elementary practice, a single, isolated magnetic pole, and the poles of any given bar magnet never appear to be situated at definite points.

Quantitative work with magnetism consists chiefly in the use of a simple magnetometer which is essentially a small permanent magnet free to rotate in a plane about its centre. The action of the instrument relies upon the fact that every magnetic dipole situated in a field experiences a couple which tends to rotate it into a position where its axis, directed from the S pole to the

N pole, lies parallel to the magnetic field. In fact, experiments with a magnetometer compare the strengths of two fields by indicating the direction of the resultant when the two fields act in perpendicular directions; when one of the fields is a standard field, e.g. the earth's, the value of the other can be deduced. Other measurements are based upon the oscillations performed by a magnet about its equilibrium position in a magnetic field, such as that of the earth.

The law governing magnetic fields is taken to be Coulomb's inverse square law between pairs of magnetic poles. Two identical poles which repel each other in a vacuum (and in the absence of any other influences) with unit force

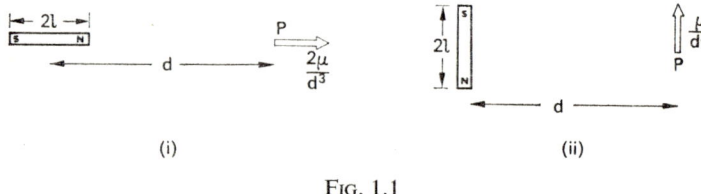

Fig. 1.1

when situated at unit distance apart are defined to be unit poles. With this definition, the force between two poles of magnitudes m_1 and m_2 at a distance d apart in a vacuum is $m_1 m_2/d^2$. A N pole has positive strength and a S pole has negative strength. (In SI the definitions of magnetic quantities follow a different sequence and start from the unit of current and the magnetic moment of a small current circuit. In contrast to the CGS electromagnetic system of units, SI does not make use of the concept of a magnetic pole as the starting point for defining magnetic quantities.)

The moment of a dipole, consisting of poles of strengths $\pm m$ separated by a distance l, is of magnitude $\mu = ml$ and is directed along the axis from the negative (S) pole to the positive (N) pole; the moment of a magnetic dipole is a vector. Using Coulomb's law and assuming that l^2/d^2 is negligible, we can show that the magnitude of the field produced by a small magnet of moment μ is $2\mu/d^3$ at a point distant d along its axis [see Fig. 1.1(i)], and is μ/d^3 at a distance d from the magnet along its bisector [see Fig. 1.1(ii)]. These are the "end-on" and "broadside" positions, sometimes called the Gauss A and B positions.

By means of a magnetometer at P, in either of the above positions, the ratio μ/H can be determined, where H is a standard field, usually the earth's field, at P acting in a direction at right-angles to the indicated field. By timing the oscillations of the magnet when suspended in the standard field at P, the product μH can be determined.

§ 1.1   INTRODUCTORY CONCEPTS   7

There are certain metals, chiefly ferrous metals, which become magnetized when they are placed in a magnetic field. When small particles of such a metal, usually "soft iron filings", are placed in a magnetic field, they tend to align themselves along the direction of the field and so form curves. The patterns so produced give a physical picture of the field traversed by *lines of force* which converge upon S poles and diverge from N poles.

One link between magnetism and electricity is epitomized in Oersted's experiment in which the presence of a current in a wire causes a deflection of a neighbouring compass needle thereby showing that a magnetic field is established in the space surrounding a conductor when a current flows in that conductor. Investigations of this magnetic field by elementary methods, including the use of iron filings, show that a single long straight current is surrounded by a field whose lines of force are circles with the centres on the conductor and planes perpendicular to the conductor. This field differs from that of any permanent magnet in that the lines of force are closed loops and do not start and finish on magnetic poles, for there are none. The direction of the lines of force is the same as that of the ends of a right-handed corkscrew when it is screwed in the direction of the current. (This is the usual right-handed relation between a sense of circulation and a direction of translation.) Other experiments indicate that a wire carrying a current and forming the perimeter of a small plane area produces a magnetic field at distant points (i.e. at points whose distance is large compared with the linear dimensions of the area) which is identical with that of a magnetic dipole. When a solenoid, consisting of a large number of turns wound closely and uniformly on a suitable former, e.g. a cylindrical one, carries a current, the magnetic field inside the solenoid is practically uniform, certainly near the centre of the solenoid, and the field outside is identical with that of a permanent magnet having the same shape as the solenoid. This magnetic effect of a current is the basis of many electrical measuring instruments, electromagnets, relays, etc. Further it provides the theoretical basis of the definition of a unit of current on the CGS system. This definition takes one of two, alternative but equivalent, forms.

(a) When a current flowing in a single circular coil of unit radius produces a magnetic field of 2π units (EM units, the oersted) at the centre, that current has unit strength.

(b) When a current flowing around the perimeter of a small plane coil enclosing area α makes the coil behave as a small magnet of moment $I\alpha$, the current has strength I (EM) units.

The unit defined thus is the (absolute) electromagnetic (EM) unit of current, sometimes called the abampere; the ampere is a current having a strength of one-tenth of this absolute unit.

Since a coil carrying a current produces a magnetic field as does a bar magnet, it is not unexpected that such a coil should experience a couple and a force when situated in a magnetic field. If a single conductor carrying a current is situated in a magnetic field, each element of the conductor experiences a force in a direction at right angles both to the conductor and to the field. The relative directions are given by Fleming's left-hand rule, viz. the directions of the force, the field and the current, taken in that order, form a *left-handed* triad. This force is illustrated in the mutual attraction between two parallel wires which carry currents in the same direction: the magnetic field of one current gives rise to the force on the other. This effect has many important applications in practice, such as electric motors of all kinds, measuring instruments, etc.

There is a second link between magnetism and electricity which was first investigated extensively by Faraday. This link is known as *electromagnetic induction*, and is quantitatively expressed in Faraday's law of electromagnetic induction. The effect occurs, in general, whenever there is a relative motion of a conductor and a magnetic field, or a change in the linkage between a circuit and the lines of force of a magnetic field. In such a case an electromotive force appears in the conductor and, if there is a suitable conducting circuit available, a current circulates in the conductor. The situations in which electromagnetic induction may occur are very varied, but whenever there is a motion of the conductor relative to the lines of the magnetic field the direction of this motion, the direction of the magnetic field, and the direction of the electromotive force form a *right-handed* triad. This is Fleming's right-hand (dynamo) rule. The consequences of this phenomenon are very far-reaching indeed: it is the fundamental principle which lies behind the operation of dynamos for generating electricity; it is the basis of the special properties of alternating currents and the action of transformers, and so plays a major role in many forms of transmission of electrical power.

5. Hertz's experiments

The experiments described thus far can all be understood in terms of a direct influence of one object (charge or current) on another across the space between them (action at a distance). Such a point of view becomes much more difficult to uphold in the experiments now to be described; and these experiments differ from the earlier ones also in being motivated by definite theoretical predictions rather than by arguments inspired by other experiments.

The theoretical predictions are due to Maxwell, and were published in 1863 (but probably known to him somewhat earlier, to judge by a notebook kept by him during his occupation of the chair in physics in King's College, London, 1860–5). Maxwell modified the electric and magnetic equations as they were known to him in a way that will be described below; and as a result he found that in free space the electric and magnetic fields both satisfied the equation of wave motion (already familiar, for water waves and sound waves) with a velocity of wave propagation determined entirely by the electric

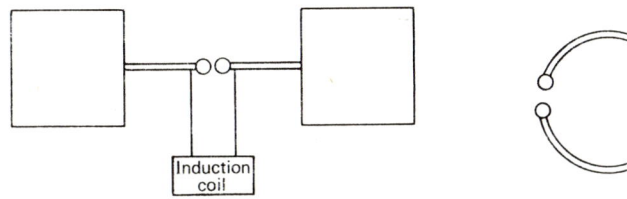

Fig. 1.2

and magnetic quantities. In fact, if ε_0, μ_0 are the constants introduced (§ 2.3, § 6.1) into the laws of electrostatics and magnetostatics respectively, the velocity is $(\varepsilon_0 \mu_0)^{-\frac{1}{2}}$. It may be noted in passing that it had been noticed by Riemann in 1853 that this velocity agreed numerically with that of light, though his note on this was not published until after his death.

Hertz published the results of his experiments in 1888. His apparatus was of the following form. Two square brass plates, each of side 40 cm, were joined by a copper wire of length 60 cm. This wire was cut in the middle and two balls fixed to make a spark-gap—that is, two surfaces across which an electric discharge is possible (Fig. 1.2). The two balls were connected to a source of electrostatic potential—in Hertz's case an "induction coil". (This consists of two coils of copper wire wound on the same iron core; one coil, the primary, has a few turns (say 300), whilst the secondary has many thousands. The primary is connected to a battery through a switch which is in turn controlled by the magnetic field set up in the iron core by the primary current, so that, as the current builds up, it is cut off again. The secondary coil is enabled to build up a high voltage by means of a condenser across it.) This part of the apparatus functioned as the transmitter of the waves.

The receiver consisted of a copper wire bent in the form of a circle of radius 35 cm, the ends almost touching and finishing with a spark-gap. When the induction coil was working sparks crossed the spark-gap of the transmitter, and the resultant *electromagnetic oscillations*, when falling on the receiver, were able to cause sparks to cross the spark-gap of the receiver. To confirm the wavelike nature of the phenomenon Hertz carried out experiments on reflection. He set his oscillator up 13 m from a wall to which a large sheet of zinc was attached. He then placed the receiver at various distances from the wall and noticed that there were periodic variations in the intensity of sparking.

This variation may be explained by supposing that the beams consist of waves, and invoking the phenomenon known as *interference* which arises with any wave motion. If the waves from a source reach the receiver in one position after following two, or more, paths of different lengths (in Hertz's experiments one path goes directly to the receiver, the other path undergoes reflection by the zinc) a crest of a wave from one path may arrive at the receiver with a trough from the other path. In this case they will cancel out and the receiver will give little response. For a different position of the receiver two crests, or two troughs, may arrive together and reinforce each other. In this case the receiver responds strongly. From the distances between positions of strong and weak response Hertz was able to deduce that the wavelength was 9·6 m. Hertz himself worked at a later date with much shorter waves, and of course the idea of electromagnetic waves has now become a commonplace, but most of the more modern applications of the idea require rather more complex application of the theory to explain them.

In view of the numerical value of the velocity of the radiation predicted by Maxwell's theory it is natural to suppose that it is of the same nature as light. Hertz was able to confirm its reflection, refraction and (more significantly) its polarization. The property of polarization was shown by the weakening of the response when the plane of the receiver was changed from vertical to horizontal, or vice versa, with the plane of the transmitter fixed. It is governed by the direction of the electric field in the radiation. (Polarization is the reason for installing TV aerials in certain areas of the country with horizontal instead of vertical branches. The transmitters in these areas send out horizontally polarized waves.) The electromagnetic theory of light envisages a complete identity between the theory of electromagnetic radiation and that of optics, so that experiments in optics are of importance for the subject-matter of this book also.

6. Atomic and Molecular Theory

Under this heading we include more modern scientific discoveries, other than those mentioned already, which have become more or less common knowledge even though the theories concerning them are not entirely understood by "the man in the street" or by the average undergraduate at the beginning of his studies.

Briefly, all forms of matter and electricity are discrete in structure. The atom consists of a positively charged nucleus, which accounts for most of the mass of the atom, surrounded by electrons moving in orbits about this nucleus. Although the picture of an atom as a kind of solar system is by no means accurate it has many of the essential features of a valid representation. All electric charges occur in integral multiples (positive or negative) of the charge carried by an electron. The electronic charge is of the kind we call negative, and a charged body, as encountered in elementary physics, has an excess or defect of electrons. The number of positive charges in an uncharged body is equal to the number of electrons. Since even a small speck of material contains a vast number of atoms, and so of both positive and negative charges, an excess of charge of one kind or another on a charged body, in the elementary experiments discussed above, is occasioned by a number of electronic charges which is vanishingly small when compared with the total number of charges in all the atoms.

The difference between a conductor and insulator lies in the relation between the electrons and the positive nuclei of the atoms. The atoms of a metal have peripheral electrons which are comparatively easily removed from the parent atoms and a piece of metal consists of a regular structure of metal atoms which have "lost contact" with their peripheral, or "conduction", electrons. These electrons, sufficient in number to keep the body as a whole electrically neutral, are more or less free to move through the interstices of the rigid structure formed by the atoms. This structure gives the metal its rigidity and crystalline nature and the presence of the "conduction" electrons, more or less free to move, gives the metal its electrical properties. On the other hand, the electrons which form part of the atoms and molecules of an insulating substance cannot easily be removed. The electrons of a solid insulator are more or less tightly bound to a fixed centre in the solid. When a conductor is placed in an electric field, the conduction electrons, being free to move, respond to the electric force by drifting, rather like a cloud, along the metal; this velocity of drift is restricted by collisions among the electrons themselves and between the electrons and the fixed framework of the metal atoms. Differences in electrical resistance correspond to the ease with which these electrons can move through the frameworks of the various substances. When an insulating substance is placed in an electric field the electrons (and positive nuclei) experience a force but are unable to move far from their equilibrium positions. They merely take up new equilibrium positions in which the electric force on them is balanced by the molecular and atomic forces binding them in position; the pattern becomes strained but the electrons do not acquire a drift velocity and no current flows.

The picture given above of the microscopic structure of a substance does not take much account of quantum mechanics. Nevertheless, if the terms "electron", "atom", etc., are not interpreted too literally as tiny billiard balls, the picture gives a general idea of the processes occurring within solids. The process of metallic conduction should be compared with electrolytic conduction: in the latter both positively and negatively charged ions move and constitute the current, the negative ions not being just electrons.

Briefly, we may say that all electric currents are charges (usually electrons) in motion. In the latter case the direction of the electrons' motion is opposite to the direction of the conventional current which is taken as the direction of motion of positive charge.

7. Other experiments

In this section we survey a number of experiments in various branches of science that have an impact on electromagnetism. These comprise only a small selection of the large number of experiments that have been performed in this century with the explicit intention of investigating the relationship between electromagnetism and the rest of physics. Our selection has been made more for the sake of the striking and simple nature of the experiments rather than for their accuracy. Indeed very much higher accuracy is possible with some of those not mentioned.

(a) Champion's experiments on electron scattering

The reader may easily verify (e.g. by using a tray with a light sprinkling of sand and two equal steel balls) that, in an elastic impact of a small sphere with another equal stationary one, the directions of the velocities of the two spheres after impact are perpendicular (Fig. 1.3).

The diagram in Fig. 1.3 is the representation of what is seen by a stationary observer in the laboratory when the moving particle has an initial velocity V. He assigns coordinates and velocities—position and velocity vectors—by a reference to a "laboratory frame" which is stationary in the laboratory. The behaviour of the particles conforms to the laws of conservation of momentum and of energy when these quantities are calculated in terms of the "laboratory frame".

Since the particles are equal, their centre of mass is always midway between them and an observer who moved so as always to coincide with the centre of mass of the two particles would see both particles moving towards him with equal and opposite velocities. He would also see them receding after the collision with equal and opposite velocities; these velocities in all cases have magnitude $\frac{1}{2}|V|$. For this observer too the laws of conservation of momentum and energy are obeyed—though the total kinetic energy in this case is $\frac{1}{4}mV^2$ and not $\frac{1}{2}mV^2$. In the view of the laboratory observer the centre of mass, and therefore the second observer's frame of reference, is in uniform motion before, during and after the collision with velocity $\frac{1}{2}V$.

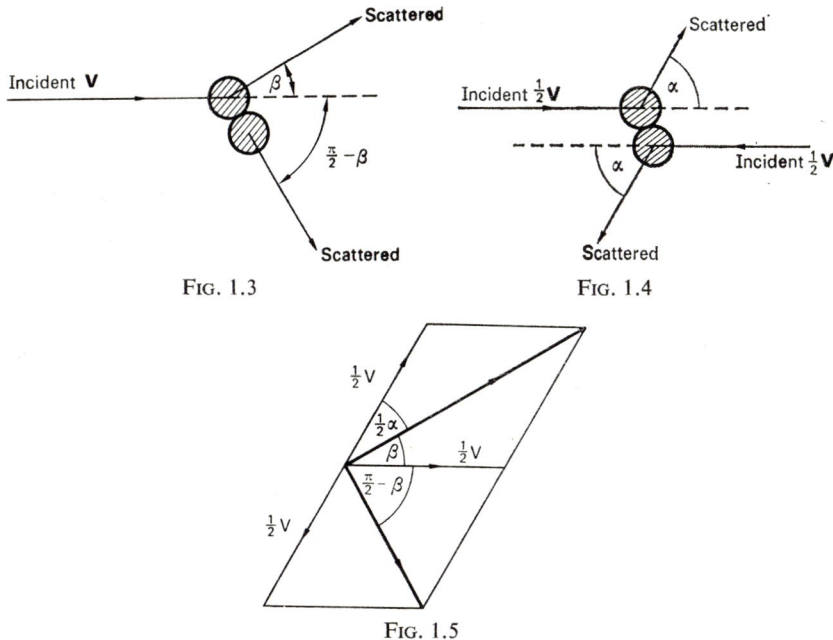

Fig. 1.3

Fig. 1.4

Fig. 1.5

If we add a velocity $\frac{1}{2}V$ to the velocities assigned by the second observer in Fig. 1.4, the "target" particle is brought to rest before the collision, and we have the state of affairs shown in Fig. 1.3. A consideration of the parallelogram of velocities concerned in adding these velocities shows that the final velocities of the incident and target particles are in directions which bisect the angle α internally and externally (Fig. 1.5). Hence, whatever the velocity V, these two particles move at right-angles after the collision in the laboratory frame. If the laboratory observer assigns position r and time t to an event, the moving observer assigns position r' and time t' to the same event where

$$r' = r - \tfrac{1}{2}Vt, \qquad t' = t,$$

i.e.

$$r = r' + \tfrac{1}{2}Vt', \qquad t = t'. \tag{1.1}$$

These equations give the transformation connecting the two frames of reference when they are in uniform relative motion $\frac{1}{2}V$. (According to the second observer the laboratory observer's frame has a velocity $-\frac{1}{2}V$.) These relations are known as *Galilean transformations*.

In 1932 Champion investigated the analogous collision between electrons at rest in a cloud chamber and incident electrons. He found that for low incident energy the tracks of the particles after scattering were at right-angles, but that at higher incident energies the tracks were closer together. Since the proof that they are at right-angles depends only on the most basic principles of mechanics (conservation of energy and momentum) and on the transformation between frames of reference in uniform motion, this suggests that some modification is needed in one of these. In fact it turns out that the transformation between two such frames in eqn. (1.1) is suspect (despite its intuitive obviousness) and needs alteration because the variable t assumed in it has not been operationally defined, i.e. no operation has been specified for measuring the time at which an event occurs, which would give the same times t, t' by both observers.

(b) The experiment of Wilson and Wilson

In 1913 A. M. and H. A. Wilson carried out an experiment designed to push to the limit the theory of electromagnetism applied to moving matter. In most problems concerning matter we have to deal with a conductor, a dielectric, or with a magnet (magnetizable) material. These forms of matter are all present in the Wilsons' experiment. The arrangement consists effectively of two conducting plates on the faces of a magnetic dielectric (in fact, wax in which a large number of steel balls are embedded) forming a condenser. This condenser (Fig. 1.6) is made to move in the direction of the x-axis through a magnetic field directed along the y-axis. The metal plates are connected by a stationary wire through carbon brushes at each end of the wire which make contact with the plates as they move.

When the velocity of the dielectric and the magnetic field are steady no current flows in the stationary wire, but when the magnetic field is reversed in direction a transient current flows in the wire until the new steady state is obtained. The current can be detected, and the total charge transferred from one plate to the other can be measured by a ballistic galvanometer.

There are in this experiment two views which can be adopted, that of the (stationary) "laboratory" observer, and that of an observer moving with the medium. The "laboratory" observer sees a magnetized medium in motion and two plates carrying charges separated by a dielectric as in a charged condenser; but because the plates are connected by a wire *there can be no electric field* between the plates, (otherwise there would be a current in the wire). Therefore the medium must be electrically polarized as a medium separating the two plates of a charged condenser is polarized. (If the medium were stationary all he would see would be a magnetized medium and no electric polarization.)

In contrast to this, the moving observer sees a stationary medium in the presence of a magnetic field and electric charges, the medium being electrically polarized and magnetized as well. The explanation of these results, quantita-

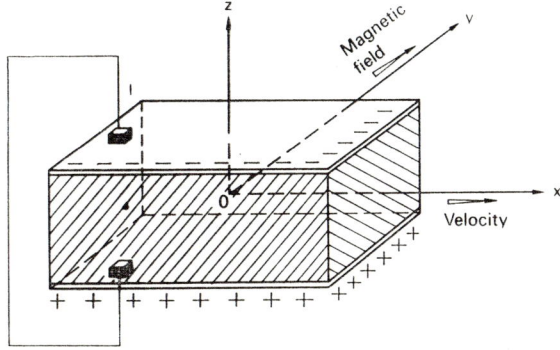

FIG. 1.6

tively as well as qualitatively can be explained only when the transformation relations between the magnetic fields and electric fields measured by observers in relative motion are known. In particular, the motion of the magnetized medium evidently has an electrical effect upon the medium.

(c) The velocity of light

The experiments of Hertz, as described in § 5, themselves suggest that some modification is needed in our formulation of the transformations between reference frames which are in relative motion. In the first place, Hertz verified a wave phenomenon of the kind predicted by Maxwell's equations, and these equations, as we said, predicted that such waves would travel with velocity $(\varepsilon_0\mu_0)^{-\frac{1}{2}}$. Notice that no mention is made here of the relative velocity of the source and the reference frame in which the observation is carried out. This is very mysterious; in the case of sound, of course, a similar equation arises, but there the velocity is measured relative to the still air. Here no mention has been made of any medium; moreover if such a medium were to exist, it would create more problems than it solves, since it would then provide from the set of inertial frames a preferred one in which Newton's first two laws of motion hold, viz. that in which the medium is locally at rest. Although Newton had in mind to find such a preferred reference frame, it would not nowadays be considered a desirable feature.

On the other hand, if we reject such a preferred coordinate system, we have to accept that the transformation between the Newtonian frames must be such that the speed of light (assuming here that it is indeed to be identified as one of the types of Hertzian radiation) has the unexpected property of not being altered when passing from one such frame to another. This is evidently *not* satisfied by the transformation

$$r \to r' = r - Vt, \qquad t' = t; \tag{1.2}$$

a fact which we had already had reason to suspect in connection with Champion's experiments.

(d) Fresnel's experiment

Some further information about the velocity of light is provided by the experiment of Fresnel, in which the speed of light in a moving medium (water) is measured. If the water is at rest it is well known that the speed of

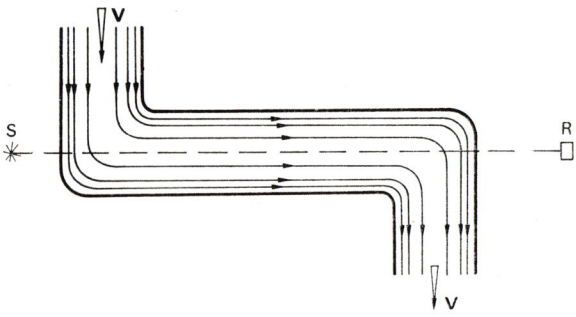

Fig. 1.7

light in it is not c but c/n, where n is its refractive index. Consider now an experiment of the type shown in Fig. 1.7, where water is shown flowing along a long pipe, with speed V. The fact that the speed of light is c/n is now true only in a frame of reference moving with the water. We ask: what is the speed measured in the laboratory system? It is to be expected that it is more than c/n; if the standard transformation [eqn. (1.2)] were employed it would be

§ 1.2   INTRODUCTORY CONCEPTS   17

$\frac{c}{n}+V$, but a moment's consideration shows this to be impossible, since the density of the fluid has not been introduced. Accordingly we could consider a sequence of experiments with rarer and rarer gases, so that $n \to 1$, and the limiting value of the observed velocity is $c+V$, in the case where nothing at all is passing along the pipe. One would expect, at least as a first approximation, that the term V would be multiplied by some factor depending on n which vanishes when $n = 1$. In fact, Fresnel, by an acoustical analogy, gave for the velocity

$$c' = \frac{c}{n} + V\left(1 - \frac{1}{n^2}\right),$$

a result confirmed experimentally by Fizeau in 1859.

These four experiments (and many others) lead us, then, to question the way in which the usually accepted transformation between frames of reference in uniform relative motion is set up; and we shall find that this is intimately connected with the relation between electric and magnetic fields.

1.2 Method and aim

We build up the theory of electromagnetism assuming a basic knowledge, summarized in § 1.1, as a background, and in our development we use mathematical techniques given in *C.M.*, vols 1–6. There are two general characteristics of this theory, which we point out now, and which permeate the whole structure.

1. MACROSCOPIC THEORY

Despite the microscopic structure of matter and electricity we attempt to build a theory which is only macroscopic in its application. Such a theory should not be applied, without modification, to microscopic systems which are known to have discontinuous properties. This implies that we shall consider charges to occur as point charges, surface charges, or volume charges, but a point charge is *not* supposed to be an electron, or any other microscopic particle; it is the counterpart in this theory of the particle in classical mechanics, viz. a charge confined within a region whose dimensions are small compared with any other length concerned in the investigation. Such an "electrical particle" is a mathematical singularity in the theory and can be approximated to a high degree of accuracy in practice by a speck of material carrying a charge. But even such a speck may contain many millions of atoms and carry a net charge which is many millions of times that carried by an electron.

The surface and volume distributions of charge which we consider are assumed to be continuous in the mathematical sense, just as laminae, shells and solid bodies are continuous distributions of mass in classical mechanics. We assume that the processes of infinitesimal calculus, which in principle involve subdivisions to an unlimited extent, are applicable to such distributions of charge.

2. Field theory

How does one charge interact with another charge at a different place and at a later time? What is the mechanism by means of which this occurs? How is it that events in the sun or distant stars affect our senses or instruments on the earth? The field outlook postulates for these processes a certain broad type of mechanism which underlies the whole theory of electromagnetism. This mechanism was mentioned first in connection with the force exerted by one current on another through the intermediary of the magnetic field. The interactions of charge with charge, current with current, etc., always take the form of a force (or couple) experienced by the object charge or current. We assume that this force is determined by some state of affairs in the space immediately surrounding the object charge or current. We say that the object is experiencing the effect of the field.

The region surrounding a charge or current has different properties from the same region in the absence of a charge or current. The difference is shown when a second charge or current experiences a force. This modified region of space is the "field" of the theory and arises from the source. We can therefore summarize this basic mechanism of an interaction postulated in the field outlook thus:

$$\text{source} \rightarrow \text{field} \rightarrow \text{force on object}.$$

If a source acts on an object in this way, the law of action and reaction requires that the object reacts on the source. We assume that the reaction occurs in a similar manner but with the roles of source and object interchanged. Hence, the designation of a particular charge or current as source or object depends upon our convenience and not upon any intrinsic differences in the system. Because of this interchangeability of source and object the field outlook regards the intervening space, the field, as the primary seat of all the phenomena—the most important link in the chain of interaction—and the source and object are taken as "end effects" of the state of affairs in the field. This field concept is so important that we illustrate the ideas by three examples drawn from theoretical mechanics. These are, of course, analogies, and must not be pressed too far.

First, consider two points A and B at the ends of an elastic string stretched beyond its natural length. An external agency must exert a force on A and another agency a force on B in order that these points shall remain stationary. Thus the agents at A and B interact, the interaction being transmitted by the string. (This corresponds to the situation of two conductors carrying charges or currents which have to be held in place by two external agents.) There are two aspects to this interaction, viz. the stretching or displacement of the string and the force or tension in the string. We can, if we wish, regard the displacement as taking place entirely at the end B, when B is moved from the "unstretched" position B_0 to B (Fig. 1.8). However, we find it more "natural" to regard the displacement as distributed along the whole length of the string, and the forces acting at A and B are the "end-effects" of a typical element at P acting on its neighbour at Q, which acts upon its neighbour, and so on. Also, each element suffers a certain displacement, the "end-effects" or sum of all

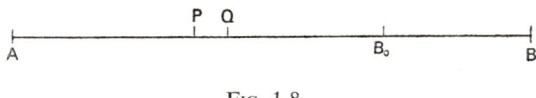

Fig. 1.8

such displacements being the displacement of B. The force required at B to hold it stationary is determined by the tension in the string in the immediate neighbourhood of B; similarly at A. When analysed in this way the string (or field) is the seat of the effects which are experienced at the ends. If we use only one string for different relative positions of A and B, we need not use both the tension and the displacement to specify the state of affairs at any point of the string. These two quantities are related by some law (Hooke's law in this case) so that specifying one fixes the other. However, if we consider the interaction transmitted by different strings, the tension corresponding to a given displacement will vary according to the string we use. Unless we know the relation between the tension and displacement we need, in general, to specify both quantities.

A second illustration, which resembles more closely an electromagnetic field in three dimensions, is provided by an inflated balloon. The balloon is inflated by injecting gas at one point A and a force is exerted on an arbitrary point B by the gas. The agent, external to the gas, which exerts the appropriate force at all points such as B, is the membrane of the balloon. If more gas is displaced into the balloon by the agent at A, then the force at B is altered: the gas transmits the action from A to B. There are two aspects of the interaction: one is the displacement of the gas, here the compression; the other is the pressure of the gas at any point.

For a third and final illustration we consider a non-uniform elastic body transmitting an interaction. A force acting at A causes the body to exert a force on the agent holding B stationary. Again, at any point of the body there are two aspects to the transmission of the stress, a force per unit area, and the strain, a measure of the displacement. These two do not necessarily act in the same direction, and the relation between them varies from one material to another.

In all three cases, string, balloon or elastic body, in order to bring about the interaction work is done by the external agencies. This work is often regarded as potential energy which is "stored" in the bodies transmitting the interaction. The above illustrations have all been cases of systems in equilibrium. We could consider also cases in which the system is not in equilibrium so that kinetic energy may also be involved. It is part of the field outlook to regard all energies as located in the region transmitting the interaction, i.e. the field, because of its primary importance.

Having used mechanical analogies above to illustrate and elucidate the concepts of a field theory, once again we warn the reader against pressing such analogies too far. In the nineteenth century electromagnetic fields were regarded as being transmitted by a medium permeating all space—the ether—which had many of the properties of an elastic solid. We do not suggest that empty space, e.g. that between the earth and the sun or stars, has any of the properties of any sort of body or material, elastic or otherwise. We use the above analogies to guide our thoughts in defining suitable quantities. No properties of free space are implied beyond those implicit in the definitions adopted and in the laws of electromagnetism which do differ from those obeyed by elastic bodies. Our aim is to build up a scheme of concepts and to postulate laws connecting the quantities we define so that the forces between bodies, particles and charges given by the theory are in agreement, so far as can be verified, with measured forces on actual bodies, etc. In addition, the objects such as charges, conductors, etc., contemplated in the theory must be realizable experimentally to a suitable degree of accuracy, in the same way that particles, rigid bodies, elastic strings, etc., can be similarly realized experimentally.

In the nineteenth century the field outlook was very important as a heuristic device, an integral part of the mechanical model-making by which the scientists of that time sought to understand electromagnetic phenomena. It can sometimes still be useful in this way, although we now realize the limited usefulness of mechanical models.

But the field concept has proved extremely useful, almost indispensable, for a new purpose. We noticed above that recent experiments have necessitated a revision of the concepts of space and time used in describing experimental results and a modification in the transformation between the results of different observers in uniform relative motion. The field concept is the mathematical means by which we are able to carry out this modification.

1.3 Scheme of development

The scheme we follow is summarized below.

First, we confine our attention to electrostatic fields in which the charges are in equilibrium, Chapters 2–4. Next, we remove the restriction to stationary charges and consider the situation when charges are in *steady motion*. This gives rise to two investigations: one concerns the behaviour of the currents themselves, how they are distributed in the conductors which carry them and the factors which govern this flow, Chapter 5; the other is the magnetic (magnetostatic) field which accompanies a steady flow of current, Chapter 6.

The next stage is to consider currents which vary, but in a restricted manner (so-called quasi-steady currents), Chapter 9. In this state of affairs the currents flow in closed circuits and at any one instant satisfy the conditions for steady currents already considered. The variations in current contemplated at this stage involve either motions of the circuits carrying the currents, or variations of the strengths of the currents in such a way that the conditions for a steady current are satisfied at any one instant. Such variations effectively imply slow changes in the current distributions from one steady state to a neighbouring steady state—hence the description "quasi-steady". Such changes in the currents imply that the magnetic field is no longer steady. It is at this point that Faraday's law of electromagnetic induction becomes relevant, i.e. variations in the magnetic field produce electric effects additional to those occasioned by the presence of stationary charges. Of course, the more general case considered here must reduce to the earlier steady-state cases when the currents are stationary.

The final stage in the development of the theory comes with the removal of all restrictions on the motion of the charges. The currents need not flow in closed circuits; together with the charges in the field the currents satisfy a *conservation law* which implies that charge moves from point to point but is not destroyed. Positive and negative charges may cancel one another but the total (net) charge in a closed system is fixed. For example, when the plates of a charged capacitor are connected by a conducting wire a current flows in the wire which is not a *closed* circuit. The current starts at one plate and ceases at the other, the conservation of charge being shown by the decrease of positive charge (or increase of negative charge) corresponding to the current flowing away from (or towards) that plate. The current in this case does not flow in a closed circuit nor does its strength remain constant as time elapses.

Because of this removal of restriction to steady currents and stationary charges, the electric field will vary. At this point the last new feature of electromagnetic theory appears, and is associated chiefly with the name of Clerk Maxwell. He postulated that variations in the electric field have magnetic effects; these variations give rise to additional magnetic effects over and above those due to conduction currents or permanent magnets. This additional source is called "displacement current", and is present whenever and wherever an electric field varies. It is the counterpart of electromagnetic induction. These two effects, the magnetic effect of a varying electric field and the electric effect of a varying magnetic field, together produce the phenomenon of electromagnetic radiation. This theoretical postulate of Maxwell, the magnetic effect of displacement current, was the chief triumph of his theory; it predicted the existence of electromagnetic radiation, a prediction which, as we have seen, was subsequently verified experimentally by Hertz.

1.4 Systems of units

We have already referred to the different sets of units which are in use for electromagnetic measurements. The system employed in this book is the SI, formerly called the MKS system, but the reader may have become acquainted with the other systems in use, the electromagnetic (EM), the electrostatic (ES), the Gaussian, and the practical systems. An important point to bear in mind is that each system, particularly the ES and EM systems, depends upon a particular sequence of definitions in addition to starting from its own basic units. Confusion arises when ES units are used to measure magnetic quantities and when EM units are used to measure electric quantities. In fact this confusion occurs where electric and magnetic phenomena are interdependent, i.e. where an electric current produces a magnetic field and in electromagnetic induction. The Gaussian set of units is not strictly a logical system of units but is a selection of EM and ES units made according to a (more or less) agreed division between electric and magnetic quantities. The practical units are prescribed multiples of the EM or ES units. The SI is a logical system of units, starting from four basic quantities. The use of a fourth basic unit enables the definitions used in the SI to furnish units such that the commonly occurring practical units are also the units given by the scheme of definitions; there is need in only a few cases to use decimal multiples of the theoretical units. It is more important for the reader to understand the logical structure of the different systems than to know the conversion factors from one system to another. The conversion factors can easily be looked up in tables, but the logical structure must be understood.

CHAPTER 2

THE INVERSE SQUARE LAW IN ELECTROSTATICS

In this chapter we define the quantities which occur in connection with electrostatic fields and show how the inverse square law can be expressed in terms of these quantities and how this law relates the electrostatic field to its sources.

2.1 Charge

As indicated in Chapter 1, we build up our scheme of measurements using the basic units of length (m), mass (kg) and time (s), together with a fourth unit which we choose to be charge. This unit is the Coulomb (C) and its magnitude is chosen as arbitrarily as are the magnitudes of the other three basic units. Probably the most convenient way of expressing this magnitude would be to give it as N positive electronic charges (cf. the modern definition of the metre as 1650 763·73 wavelengths of a certain line in the spectrum of krypton.) Such a definition would not give an easily reproducible practical means of obtaining one Coulomb, so, in fact, the specification is given by means of the uniform current which corresponds to the passage of one coulomb in one second. The uniform current can be determined by the force between two suitably arranged conductors each carrying that current. However, for our purposes we take the coulomb as a fourth basic unit to be a positive charge of magnitude $6\cdot 242\times 10^{18}$ times the electronic charge. Once we have a unit charge, an arrangement such as an electroscope and Faraday's ice-pail enables us, in principle, to determine any rational multiple or submultiple of that unit charge.

We assume that a charge of known magnitude may be concentrated at a point, spread continuously on a line, spread continuously on a surface, or fill a volume with a continuous distribution of charge.

(1) We denote a point charge by Q.

(2) If an arbitrary line element of length l carries a charge Q, we define the line density of charge to be
$$e = \lim_{l \to 0} (Q/l). \tag{2.1}$$

(3) If an arbitrary element of area α, carrying a surface charge, is associated with a charge Q, we define the surface density of the charge to be
$$\sigma = \lim_{\alpha \to 0} (Q/\alpha), \tag{2.2}$$

i.e. the charge on an element of area α is $\sigma\alpha$, correct to the first order in α, when α is small.

(4) Similarly, if a volume v of a continuous volume distribution holds a charge Q, then the volume density of charge is defined by
$$\varrho = \lim_{v \to 0} (Q/v). \tag{2.3}$$

This means that, correct to the first order in v, a small volume v of the distribution holds a charge ϱv. Both ϱ, σ are scalar functions of position.

2.2 The electric field and electric potential

Since we adopt the field outlook in analysing the interaction of two charges we next *define* the *field strength* (or *electrostatic intensity*) corresponding to the object end of the interaction chain. The existence of a field is demonstrated when a charge experiences a force, this force being determined by the field in the immediate vicinity of the charge. We therefore imagine a charge, which we call a "test" charge, placed at a point in the field; when the test charge Q experiences a force F we define the field strength to be
$$\boldsymbol{E} = \lim_{Q \to 0} (F/Q). \tag{2.4}$$

We use the limit in the definition to ensure that the presence of the test charge does not disturb the field. (The currents in a river can be investigated by the forces acting on small material particles in the water, but not by the forces acting on a large ship—the ship interferes with the currents being investigated. Similarly a large test charge affects the field.) The usual form of eqn. (2.4) states that the force F acting on a point charge Q situated at a point where the field strength is $\boldsymbol{E}$ is given by
$$\boldsymbol{F} = Q\boldsymbol{E}.$$

If a test charge [here and subsequently the phrase "test charge" implies the limiting process of eqn. (2.4)] is moved along an arbitrary path in the field, so that it returns to its starting point, and the positions of charges, conductors, etc., in the field remain unchanged, the principle of conservation of energy requires that no net work is done during the circuit by the force exerted by the field, or, equivalently, by the agent moving the test charge. This, expressed symbolically, gives

$$\lim_{Q \to 0} \left\{ \frac{1}{Q} \oint \mathbf{F} \cdot \mathbf{ds} \right\} = \oint \mathbf{E} \cdot \mathbf{ds} = 0 \tag{2.5}$$

for an arbitrary closed contour, and implies that

$$\mathbf{curl\ E} = \mathbf{0}. \tag{2.6}$$

We can therefore find a scalar function V such that

$$\mathbf{E} = -\mathbf{grad}\ V. \tag{2.7}$$

(Note the minus sign, which is conventional.) Provided that the region in which eqn. (2.6) holds is singly connected, the function V is single-valued. If the region has p-fold connection, V is a cyclic function with $p-1$ cyclic constants. [See *C.M.*, vol. 4, § 1.5 (5).] As defined by eqn. (2.7) the function V is indeterminate to the extent of adding an arbitrary constant. The definition adopted for V makes it unique by writing

$$V(P) = \int_P^\infty \mathbf{E} \cdot \mathbf{ds} = -\int_\infty^P \mathbf{E} \cdot \mathbf{ds}, \tag{2.8}$$

which effectively postulates that V is zero at infinity. Expressed in words this definition is:

> "The potential at P is the work done, per unit test charge, by the force of the field when the test charge moves from P to infinity."

This leads to the first integral in eqn. (2.8)

The concept of *field lines* was mentioned in Chapter 1 in connection with a magnetic field; a similar concept can be used in connection with any vector field. A *line of force* of an electrostatic field is a curve drawn in the field such that the tangent to this curve at any point is in the direction of the field vector $\mathbf{E}$ at that point. Except where $\mathbf{E} = \mathbf{0}$, a line of force must have a unique direction and no two lines of force can cross. The differential equations of a line of force, by its definition, are

$$\frac{dx}{E_x} = \frac{dy}{E_y} = \frac{dz}{E_z},$$

in cartesian coordinates. These equations can be formulated in any convenient system of coordinates, e.g. in spherical polars they are

$$\frac{dr}{E_r} = \frac{r\,d\theta}{E_\theta} = \frac{r\sin\theta\,d\psi}{E_\psi},$$

where E_r, E_θ, E_ψ are the radial, longitudinal and transverse resolutes of E. (See *C.M.*, vol. 4, eqn. (4.2).]

Further, we can draw surfaces in the field on which V is constant; these are equipotential surfaces. Equation (2.7) shows that the lines of force are the orthogonal trajectories of the equipotential surfaces and the minus sign in eqn. (2.7) implies that the positive sense of the line of force corresponds to the direction of falling potential. In Figs. 2.1(i)–(vi) we sketch the rough shapes of the lines of force of some simple electrostatic fields.

2.3 Electric displacement

In terms of the analogies used in Chapter 1 the field vector E corresponds to the force or tension in the string, the pressure of the gas, or the stress in the elastic body. We turn now to the other aspect, the displacement. (Once more we warn the reader against pressing any analogy too far.)

Each of Figs 2.1(i)–(vi) represents a field as a region threaded by lines of force which start from a positive charge and finish on an equal negative charge. This suggests that, in a sense, charge has been transferred along the direction of the lines of force. Following the analogies above we say that there is a "displacement" of charge at every point of the field and that the positive and negative charges at the boundaries are the end-effects of this displacement. We define a second vector field D, the *electric displacement*, as follows

$$\boldsymbol{D}\cdot\hat{\boldsymbol{e}} = \lim_{Q,\alpha\to 0}(Q/\alpha), \tag{2.9}$$

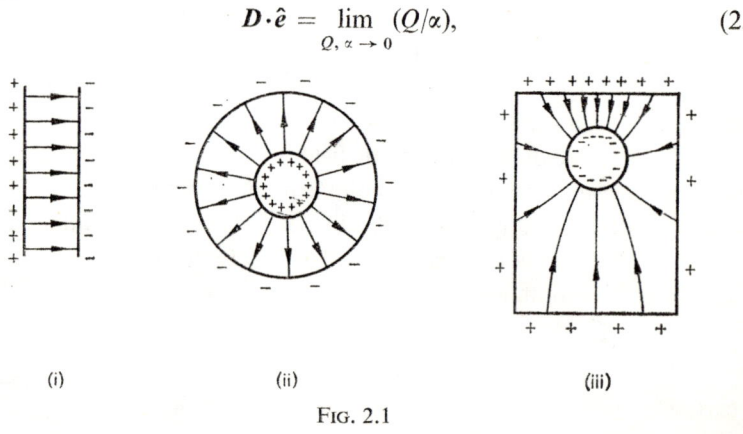

Fig. 2.1

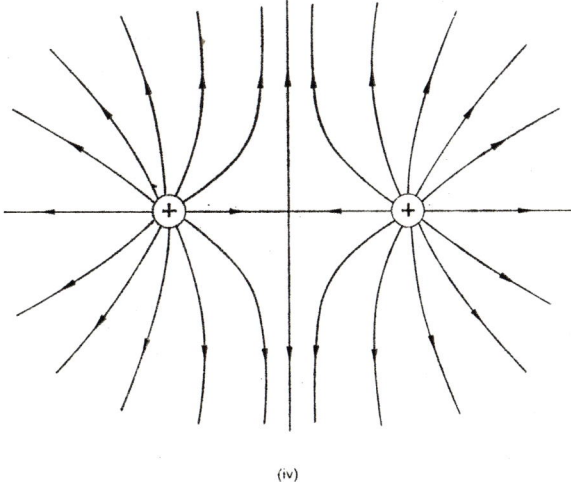

(iv)

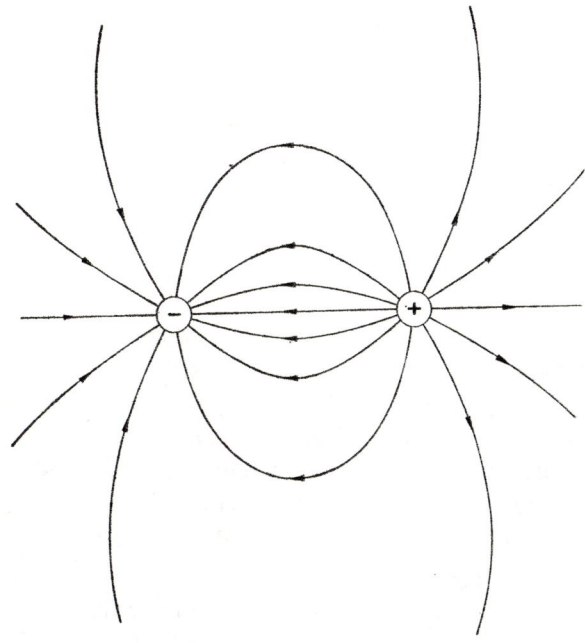

(v)

Fig. 2.1 *(cont.)*

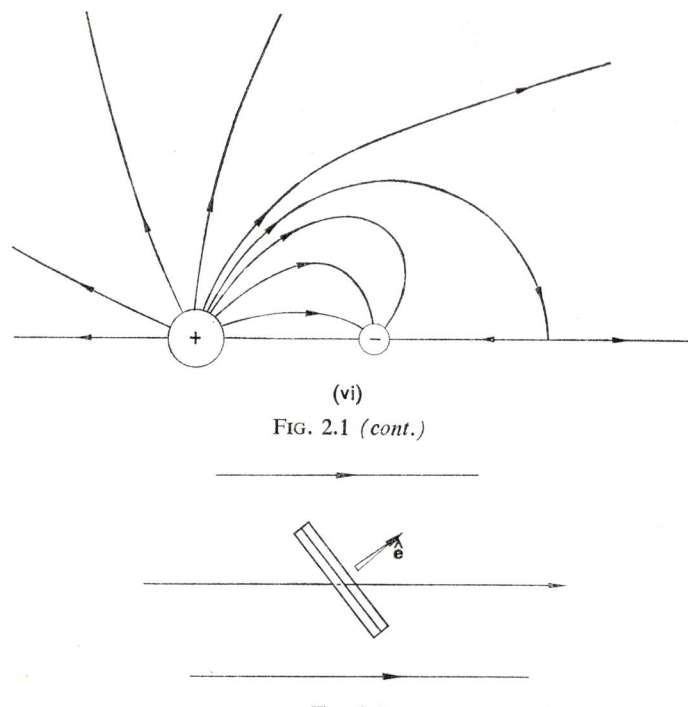

(vi)

Fig. 2.1 (cont.)

Fig. 2.2

where Q is the charge displacement across a small (vector) area $\alpha\hat{e}$ at an arbitrary point of the field, the direction $\hat{e}$ also being arbitrary. The resolutes of D are determined by taking $\hat{e}$ parallel to the coordinate directions in succession. A direct but approximate "experimental" determination of D can be made by taking two small, thin, identical metal discs of area α and holding them, in contact, back to back in the field (Fig. 2.2). Whilst still in the field the discs are separated so that they are insulated from each other, and then removed from the field. The plate on the positive side, given by the direction of $\hat{e}$, will carry a charge $\alpha(D\cdot\hat{e})$ and the other plate will carry a charge $-\alpha(D\cdot\hat{e})$. Since the plates are conductors, when they are inserted in contact in the field the electric displacement in the vacuum is replaced by an actual separation of the charges on the conductor by the process of influence.

We have now defined two field vectors E, D corresponding to the two aspects of an interaction transmitted by a field. If we were to confine our attention to a vacuum, we should need only one of these vectors to describe the state of affairs adequately at any point of the field. However, when we come to consider different media the differences between the media will be shown by differences in the relation between the two vectors E, D. The vacuum being

§ 2.4   THE INVERSE SQUARE LAW IN ELECTROSTATICS   29

the simplest possible "medium", the relation, *in this case*, is taken to be

$$D = \varepsilon_0 E. \tag{2.10}$$

The constant ε_0 is called the *permittivity of a vacuum*, its value being a universal constant.

Before proceeding further we give here a list of the names of the units defined thus far. Some of the units, those in more frequent use, have names of their own, others have compound names for their units (cf. velocity in ms^{-1}, or acceleration in ms^{-2}). The four basic units and the mechanical units constructed from them are, together with the accepted abbreviations:

length : metre,	m,	force : newton,	N,
mass : kilogramme,	kg,	work : joule,	J,
time : second,	s,	power : watt,	W.
charge : coulomb,	C,		

The definition of E, eqn. (2.4), gives its unit as newton per coulomb (NC^{-1}). However, potential (difference) has the volt as its unit, and so from eqn. (2.7) we have the alternative name of volt per metre (Vm^{-1}), which is more commonly used for electric field. [The volt itself, being defined by $P = VI$, is the same as one watt per ampere (WA^{-1}), or one joule per coulomb (JC^{-1}).]

The definition of D in eqn. (2.9) shows that its unit is coulomb per square metre (Cm^{-2}), the same as surface density of charge.

Finally, eqn. (2.10) shows that the unit of permittivity is that of displacement divided by electric field, giving coulomb per volt-metre as its unit. But we shall see later that capacitance is measured by coulombs per volt, a unit which is given the name farad (F); hence the unit of permittivity is usually taken as a farad per metre (Fm^{-1}).

2.4 Coulomb's inverse square law

The model of the electrostatic field constructed in the preceding sections depends essentially on the idea that the space (vacuum) surrounding a number of charges is modified because of the presence of these charges, the modification being described by two field vectors E, D, which are related by $D = \varepsilon_0 E$ (for the simple case of a vacuum). We now consider how D at any point is related to the charges which are the source of the field. Since $E = D/\varepsilon_0$, this relation determines the force exerted on a charge at any point by the sources of the field.

Coulomb's law gives the force between two point charges, *in the absence of any other charges*; it states that:

The force exerted by one point charge on another is directly proportional to the product of the magnitudes of the charges, is inversely proportional to the square of the distance between them, acts along the line joining them, and is a repulsion when the charges have the same sign, and is an attraction when they have opposite signs.

We can represent this by the relation

$$F \sim \frac{Q_1 Q_2}{r^3} r, \qquad (2.11)$$

where F is the force between the charges, Q_1 and Q_2 are the strengths of the charges, r is the position vector of the charge experiencing the force referred to the other charge as origin, and $r = |r|$.

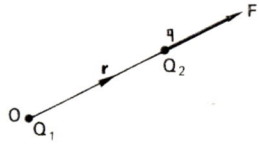

Fig. 2.3

This law was verified approximately by direct measurement by Coulomb (hence the name) and also by Cavendish. It has been verified subsequently to a high degree of accuracy, directly and indirectly by a variety of experiments.

Coulomb's law enables us to relate the vector D to the sources of a field and hence to obtain the complete chain of interaction between the sources of a field and an object charge. In the situation contemplated by Coulomb's law we regard one of the charges, say Q_1, as the source of the field and we regard the other, Q_2, as a test charge which experiences the force but does not disturb the field. We choose Q_1 as the origin of coordinates (Fig. 2.3). We may write eqn. (2.11) as

$$F = k \frac{Q_1 Q_2}{r^3} r,$$

where k is a constant of proportionality. Therefore

$$E = \frac{F}{Q_2} = k \frac{Q_1}{r^3} r, \quad D = \varepsilon_0 k \frac{Q_1}{r^3} r.$$

This result shows that the vector D (and E) arising from Q_1 has spherical symmetry, the lines of force being everywhere radial. These lines arise from

§ 2.4 THE INVERSE SQUARE LAW IN ELECTROSTATICS

Q_1 at O and must end on $-Q_1$ dispersed at infinity, so that the total displacement across a sphere of radius r centred on O is Q_1. Therefore

$$Q_1 = 4\pi r^2 |\boldsymbol{D}| = 4\pi r^2 D_r = 4\pi\varepsilon_0 k Q_1,$$

$$k = \frac{1}{4\pi\varepsilon_0}, \quad F = \frac{Q_1 Q_2}{4\pi\varepsilon_0 r^3} r,$$

$$\boldsymbol{D} = \frac{Q_1}{4\pi r^3} \boldsymbol{r} = \varepsilon_0 \boldsymbol{E}. \tag{2.12}$$

Equation (2.12) gives the displacement at any point produced by a point charge Q_1 at the origin. The potential function of this spherically symmetrical field must be a function of r only. Using spherical polars

$$\boldsymbol{E} = \{E_r \ 0 \ 0\}, \quad \text{grad } V = \left\{\frac{dV}{dr} \ 0 \ 0\right\},$$

$$E_r = -\frac{dV}{dr} = \frac{Q_1}{4\pi\varepsilon_0 r^2}, \quad V = V_0 + \frac{Q_1}{4\pi\varepsilon_0 r}.$$

If we impose the condition that $V \to 0$ as $r \to \infty$ [see eqn. (2.8)], we deduce that $V_0 = 0$, i.e.

$$V = \frac{Q_1}{4\pi\varepsilon_0 r}. \tag{2.13}$$

We now generalize this law beyond the limits for which it has been directly verified. To do this we make two assumptions on which we base our further development. The assumptions are:

(1) We assume that if one charge distribution produces a field $\boldsymbol{E}_1$ at an arbitrary point, and that if a second charge distribution produces a field $\boldsymbol{E}_2$ at the same point, then the two distributions when present together produce a field $\boldsymbol{E} = \boldsymbol{E}_1 + \boldsymbol{E}_2$ at that point, i.e. we assume that

the field vectors, $\boldsymbol{E}$, $\boldsymbol{D}$ are linear functions of the charges producing the field.

This is sometimes called the *principle of superposition*.

(2) We also assume that every point charge, and every infinitesimal element of continuous distributions, produces a field at an arbitrary point according to Coulomb's law, the resultant field being the vector sum of the contributions from every charge in the field, i.e. we assume that *every element of charge which is the source of a field contributes to that field according to Coulomb's law.*

When, in addition to point charges, continuous volume and surface charges are present the above assumptions lead to the following expressions:

$$D = \iiint \frac{\varrho r}{4\pi r^3}\, d\tau + \iint \frac{\sigma r}{4\pi r^3}\, dS + \sum_l \frac{Q_l r}{4\pi r^3} = \varepsilon_0 E, \qquad (2.14)$$

$$V = \iiint \frac{\varrho}{4\pi\varepsilon_0 r}\, d\tau + \iint \frac{\sigma}{4\pi\varepsilon_0 r}\, dS + \sum_l \frac{Q_l}{4\pi\varepsilon_0 r}. \qquad (2.15)$$

The interpretation of the quantities in eqns. (2.14) and (2.15) is shown in Fig. 2.4. The field point P has coordinates $\{x\, y\, z\}$ and the source point has coordi-

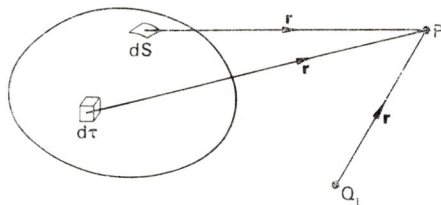

Fig. 2.4

nates $\{\xi, \eta, \zeta\}$. The sources may be either a volume element of charge $\varrho\, d\tau$, a surface element of charge $\sigma\, dS$, or a point charge Q_l (the suffix l numbers the point charge sources). In each case the vector r is drawn from the source point to the field point, i.e. in cartesian coordinates

$$r = i(x-\xi) + j(y-\eta) + k(z-\zeta), \quad r = |r|.$$

In eqns. (2.14) and (2.15) the (dummy) variables of integration are $\{\xi\, \eta\, \zeta\}$ and the sum denoted by $\sum_l$ includes all point charges which are sources of the field. In denoting the field at an arbitrary field point P we usually use the plain symbols V, D, E, but where we wish to emphasize the field point P we use symbols such as $V(P), D(P), E(P)$ or V_P, D_P, E_P, etc. We now give some examples of the use of formulae (2.14) and (2.15).

Example 1. We obtain expressions for the field quantities at P, $\{x\, y\, z\}$, due to point charges Q_l situated on the z-axis at the points $\{0\ 0\ z_l\}$, $(l = 1, 2, \ldots, N)$. (See Fig. 2.5.)

$$D = i \sum_l \frac{Q_l x}{4\pi r_l^3} + j \sum_l \frac{Q_l y}{4\pi r_l^3} + k \sum_l \frac{Q_l(z-z_l)}{4\pi r_l^3} = \varepsilon_0 E,$$

$$V = \sum_l \frac{Q_l}{4\pi\varepsilon_0 r_l}.$$

This field clearly has cylindrical symmetry about the z-axis.

§ 2.4 THE INVERSE SQUARE LAW IN ELECTROSTATICS

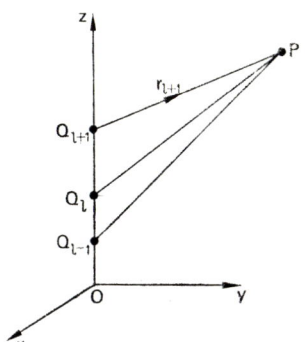

Fig. 2.5

Example 2. The field quantities due to a charge distributed along the z-axis between A, $\{0\ 0\ a\}$, and B, $\{0\ 0\ b\}$, with uniform line density (charge per unit length) e. (See Fig. 2.6 in which PN is perpendicular to Oz.)

$$D_x = \int_a^b \frac{ex\, d\zeta}{4\pi\{x^2+y^2+(z-\zeta)^2\}^{3/2}} = \varepsilon_0 E_x,$$

$$D_y = \int_a^b \frac{ey\, d\zeta}{4\pi\{x^2+y^2+(z-\zeta)^2\}^{3/2}} = \varepsilon_0 E_y,$$

$$D_z = \int_a^b \frac{e(z-\zeta)\, d\zeta}{4\pi\{x^2+y^2+(z-\zeta)^2\}^{3/2}} = \varepsilon_0 E_z,$$

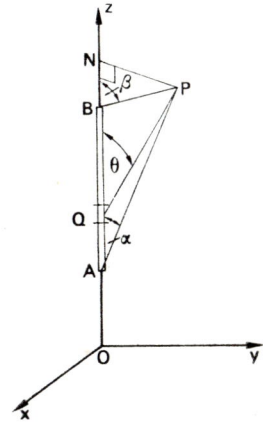

Fig. 2.6

where $r^2 = x^2+y^2+(z-\zeta)^2$. Thes integrals are best evaluated by the substitution which makes the angle θ the variable of integration, viz.

$$p = PN = r\sin\theta, \quad QN = z-\zeta = r\cos\theta,$$

so that
$$z-\zeta = p\cot\theta, \quad d\zeta = p\,\mathrm{cosec}^2\theta\, d\theta.$$

Hence

$$D_x = \frac{e}{4\pi} \int_\alpha^\beta \frac{xp \operatorname{cosec}^2 \theta \, d\theta}{p^3 \operatorname{cosec}^3 \theta} = \frac{ex}{4\pi p^2} \int_\alpha^\beta \sin\theta \, d\theta = \frac{ex}{4\pi p^2} (\cos\beta - \cos\alpha),$$

where θ takes the values α, β at A, B respectively. Similarly

$$D_y = \frac{ey}{4\pi p^2} (\cos\beta - \cos\alpha),$$

$$D_z = \frac{e}{4\pi} \int_\alpha^\beta \frac{p^2 \cot\theta \operatorname{cosec}^2 \theta \, d\theta}{p^3 \operatorname{cosec}^3 \theta} = \frac{e}{4\pi p} \int_\alpha^\beta \cos\theta \, d\theta = \frac{e}{4\pi p} (\sin\alpha - \sin\beta).$$

This is a field with cylindrical symmetry and, using cylindrical polar coordinates $\{\tilde{\omega} \; \phi \; z\}$ to give the position of P, $x = \tilde{\omega} \cos\phi$, $y = \tilde{\omega} \sin\phi$, leading to

$$D_{\tilde{\omega}} = \frac{e}{4\pi\tilde{\omega}} (\cos\beta - \cos\alpha), \quad D_\phi = 0, \quad D_z = \frac{e}{4\pi\tilde{\omega}} (\sin\alpha - \sin\beta).$$

The potential is given by the integral

$$V = \frac{e}{4\pi\varepsilon_0} \int_a^b \frac{d\zeta}{\{x^2 + y^2 + (z-\zeta)^2\}^{1/2}} = \frac{e}{4\pi\varepsilon_0} \int_\alpha^\beta \frac{p \operatorname{cosec}^2 \theta \, d\theta}{p \operatorname{cosec}\theta}$$

$$= \frac{e}{4\pi\varepsilon_0} \int_\alpha^\beta \operatorname{cosec}\theta \, d\theta = \frac{e}{4\pi\varepsilon_0} \Big[-\ln(\operatorname{cosec}\theta + \cot\theta) \Big]_\alpha^\beta$$

$$= \frac{e}{4\pi\varepsilon_0} \ln \left| \frac{\operatorname{cosec}\alpha + \cot\alpha}{\operatorname{cosec}\beta + \cot\beta} \right|.$$

The result can be put into a more useful form by writing

$$\left| \frac{\operatorname{cosec}\alpha + \cot\alpha}{\operatorname{cosec}\beta + \cot\beta} \right| = \left| \frac{PA + AN}{PB + BN} \right| = \left| \frac{PA + AB + BN}{PB + BN} \right|.$$

But $PA^2 = AB^2 + PB^2 + 2AB \cdot BN$, and substituting for BN gives

$$\left| \frac{\operatorname{cosec}\alpha + \cot\alpha}{\operatorname{cosec}\beta + \cot\beta} \right| = \left| \frac{2(PA + AB) \cdot AB + PA^2 - AB^2 - PB^2}{2PB \cdot AB + PA^2 - AB^2 - PB^2} \right|$$

$$= \left| \frac{(AB + PA)^2 - BP^2}{PA^2 - (PB - AB)^2} \right| = \left| \frac{PA + PB + AB}{PA + PB - AB} \right|.$$

Therefore
$$V = \frac{e}{4\pi\varepsilon_0} \ln \left| \frac{PA + PB + AB}{PA + PB - AB} \right|.$$

We see, from this result, that the equipotential surfaces are such that $PA + PB = \text{const.}$, i.e. they are ellipsoids of revolution having A, B as foci. Also the field vectors at P, being perpendicular to the equipotential surface, bisect the angle APB.

Example 3. We find the field quantities at a point $\{0 \; 0 \; z\}$ on the axis of a disc of charge, the surface density of the charge being constant, σ, over the area $x^2 + y^2 < a^2$, $z = 0$.

We divide the disc into elements of area, using polar coordinates in the plane of the disc, the typical element carrying a charge $\sigma r \, dr \, d\theta$, Fig. 2.7. Then

$$D_x = \iint \frac{\sigma(-x) r \, dr \, d\theta}{4\pi (r^2 + z^2)^{3/2}} = -\frac{\sigma}{4\pi} \iint \frac{r^2 \cos\theta \, dr \, d\theta}{4\pi (r^2 + z^2)^{3/2}}.$$

The integral is taken over the area of the disc. The θ-integration gives

$$D_x = 0 = \varepsilon_0 E_x.$$

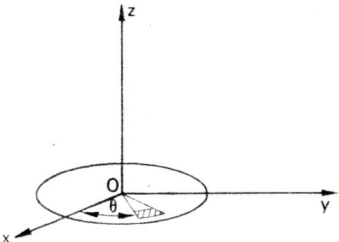

Fig. 2.7

Similarly $D_y = 0 = \varepsilon_0 E_y$. These two results are to be expected from the symmetry of the field. Also, for $z > 0$,

$$D_z = \iint \frac{\sigma z r \, dr \, d\theta}{4\pi(r^2+z^2)^{3/2}} = \frac{\sigma}{4\pi} \int_0^{2\pi} d\theta \int_0^a \frac{zr \, dr}{(r^2+z^2)^{3/2}}$$

$$= \frac{\sigma}{2} \left[-\frac{z}{\sqrt{(r^2+z^2)}} \right]_{r=0}^{r=a} = \frac{\sigma}{2} \left(1 - \frac{z}{\sqrt{(a^2+z^2)}} \right).$$

In vector terms this result is, for $z > 0$,

$$\mathbf{D} = \frac{1}{2} \sigma \mathbf{k} \left(1 - \frac{z}{\sqrt{(a^2+z^2)}} \right), \tag{1}$$

and for $z < 0$,

$$\mathbf{D} = \frac{1}{2} \sigma(-\mathbf{k}) \left(1 + \frac{z}{\sqrt{(a^2+z^2)}} \right). \tag{2}$$

If we denote by $\mathbf{D}_+, \mathbf{D}_-$ respectively the value of $\mathbf{D}$ on the positive and negative sides of the disc, as $z \to 0$ we obtain from eqns. (1) and (2)

$$\mathbf{D}_+ - \mathbf{D}_- \to \sigma \mathbf{k}. \tag{3}$$

The potential is

$$V = \iint \frac{\sigma r \, dr \, d\theta}{4\pi\varepsilon_0 \sqrt{(r^2+z^2)}} = \frac{\sigma}{2\varepsilon_0} \left[\sqrt{(r^2+z^2)} \right]_{r=0}^{r=a} = \frac{\sigma}{2\varepsilon_0} \{ \sqrt{(a^2+z^2)} - |z| \}.$$

Using a similar notation to that in eqn. (3) we see that, as $z \to 0$, $V \to \sigma a^2/(2\varepsilon_0)$ and that $V_+ - V_- \to 0$.

We emphasize that these results apply *only* to field points on the axis. For field points off the axis, or for discs which are not circular, we cannot deduce that $\mathbf{D}$ is in the direction of the normal $\mathbf{k}$.

Exercises 2.4

1. A charge e is uniformly distributed on the circle with equations $x^2 + y^2 = a^2$, $z = 0$. Show that the potential at the point $\{0 \ 0 \ z\}$ is $e/\{4\pi\varepsilon_0 \sqrt{(a^2+z^2)}\}$ and that the electric field there is $ez\mathbf{k}/\{4\pi\varepsilon_0(a^2+z^2)^{3/2}\}$.

2. The plane region inside a circle of radius a carries a charge distribution of surface density $1 - 2r^2$, where r is the distance from the centre. Prove that, if $a > 1$, there is a neutral point on the axis of the circle at a distance $1/\{2\sqrt{(a^2-1)}\}$ from the centre of the circle.

2.5 Two-dimensional fields

Two-dimensional fields are characterized by the fact that the electric field is everywhere parallel to a fixed plane. Such fields arise from special distributions of charge, which are uniform in one direction. For example, an infinite straight line of charge, of uniform line density, is one such distribution; an infinite circular cylinder carrying a uniform surface density is another. Any charge distribution such that all its sections perpendicular to the one given direction are identical gives rise to a field which has no component in the uniform direction, i.e. a two-dimensional field.

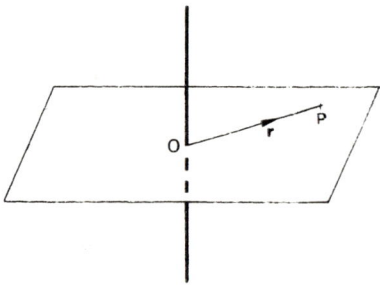

Fig. 2.8

To investigate such fields we need consider only a typical section perpendicular to the uniform direction; each part of the charge distribution is replaced by its trace on this plane, e.g. an infinite line is replaced by a point, an infinite circular cylinder by a circle. The charge distribution is specified by the charge on unit length of the uniform direction. Thus a line charge of density e per unit length is represented by a "point" charge e in the two-dimensional field; a circular cylinder carrying a uniformly distributed charge e on unit length is represented by a circle around which the charge e is spread with a density $e/(2\pi a)$.

First we find the field quantities corresponding to an infinite charge of line density e. This line charge cuts a typical section in a point O, Fig. 2.8, and we choose a field point P anywhere in this plane, taking the position vector to be $r = \overrightarrow{OP}$. To find the field quantities we take the results of example 2, p. 33, and increase the length of the segment AB so that $\alpha \to 0, \beta \to \pi$. Then

$$D_x \to \frac{ex}{2\pi r^2}, \quad D_y \to \frac{ey}{2\pi r^2}, \quad D_z \to 0.$$

Therefore
$$D = \frac{e}{2\pi r^2} r = \varepsilon_0 E. \tag{2.16}$$

§ 2.5 THE INVERSE SQUARE LAW IN ELECTROSTATICS

To find the limiting value of V, we write $\beta = \pi - \beta'$ and take the limit, obtaining

$$V = \lim \frac{e}{4\pi\varepsilon_0} \ln \left| \cot\left(\frac{1}{2}\alpha\right) \cot\left(\frac{1}{2}\beta'\right) \right| = -\lim \frac{e}{4\pi\varepsilon_0} \ln \left| \tan\frac{1}{2}\alpha \tan\frac{1}{2}\beta' \right|,$$

as $\alpha \to 0$, $\beta' \to 0$, $V \to \infty$. However, we qualify this result as follows.

When α, β' are small $\tan\frac{1}{2}\alpha \approx \frac{1}{2}\alpha \approx \frac{r}{2NA}$, $\tan\frac{1}{2}\beta' \approx \frac{1}{2}\beta' \approx \frac{r}{2NB}$.
[Here we use r in place of $\tilde{\omega}$.]

Therefore $V \approx -\dfrac{e}{4\pi\varepsilon_0} \ln\left(\dfrac{r}{2NA} \cdot \dfrac{r}{2NB}\right) = -\dfrac{e}{2\pi\varepsilon_0} \ln r + \dfrac{e}{2\pi\varepsilon_0} \ln (4NA \cdot NB),$

so that in the limit

$$V = V_0 - \frac{e}{2\pi\varepsilon_0} \ln r. \tag{2.17}$$

In the limiting process NA, NB both tend to infinity but r remains finite as the length AB is increased indefinitely; hence, the infinite value of V arises from the term V_0 which is independent of r, i.e. an infinite constant. On differentiating V the (infinite) constant V_0 disappears giving the result (2.16). The difficulty of the infinite constant arises because the charge distribution being considered cannot be enclosed in any surface which is everywhere at a finite distance from the origin. In practice, the constant V_0 is ignored, or omitted, because it does not affect the field vectors, and we take the two-dimensional field arising from an infinite line charge e per unit length to be given by

$$\mathbf{D} = \frac{e}{2\pi r^2}\mathbf{r} = \varepsilon_0 \mathbf{E}, \quad V = -\frac{e}{2\pi\varepsilon_0} \ln r, \quad |\mathbf{r}| = r. \tag{2.18}$$

All the vectors lie in a typical plane perpendicular to the line of charge. Equation (2.18) gives the form taken by Coulomb's law for a two-dimensional field. For field points close to the centre of a line charge of finite length eqn. (2.18) gives the approximate values for the field quantities. The approximation is valid when the distance r is small compared with the length of the line of charge, when the ellipsoidal equipotential surfaces can be taken as cylindrical over a short zone near their "equators".

Example. A number of infinite line charges are represented on a typical section by the points $A_1, A_2, A_3, \ldots$, the respective line densities being $e_1, e_2, e_3, \ldots$ (Fig. 2.9). The field quantities at $P, \{x\, y\}$, are

$$\mathbf{D} = \mathbf{i} \sum_l \frac{e_l(x-\xi_l)}{2\pi r_l^2} + \mathbf{j} \sum_l \frac{e_l(y-\eta_l)}{2\pi r_l^2} = \varepsilon_0 \mathbf{E}, \tag{1}$$

$$V = V_0 - \frac{1}{2\pi\varepsilon_0} \ln (r_1^{e_1} r_2^{e_2} r_3^{e_3} \ldots). \tag{2}$$

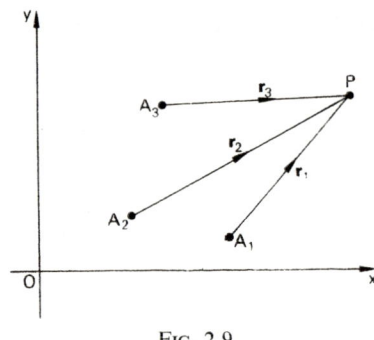

Fig. 2.9

In practice formulae such as (1) and (2) are seldom used for two-dimensional fields; a different, more powerful method, based upon the theory of functions of a complex variable is used for such fields. This method is considered later in Vol. 2, Chap 8, § 8.4. See also Exercises 2.7, question 1.

2.6 The field at internal points

In the expressions (2.14) and (2.15) for D, E and V the field point P has coordinates $\{x\,y\,z\}$. Provided that P does not coincide with one of the point charges Q_l, does not lie on a surface carrying a surface density of charge σ, and does not lie inside the region containing volume charge ϱ, all these integrals are continuous functions of the position of P and can be differentiated under the integral signs w.r. to the coordinates x, y, z any number of times. Hence we obtain

$$-\mathbf{grad}\,V = -\iiint \frac{\varrho}{4\pi\varepsilon_0}\,\mathbf{grad}_P\left(\frac{1}{r}\right)\,d\tau - \iint \frac{\sigma}{4\pi\varepsilon_0}\,\mathbf{grad}_P\left(\frac{1}{r}\right)\,dS$$
$$-\sum_l \frac{Q_l}{4\pi\varepsilon_0}\,\mathbf{grad}_P\left(\frac{1}{r}\right),$$

where the symbol $\mathbf{grad}_P$ signifies that the differentiations are performed w.r. to the coordinates $\{x\,y\,z\}$ of P. Since $\mathbf{grad}_P\left(\dfrac{1}{r}\right) = -\dfrac{\mathbf{r}}{r^3}$, we see that

$$\mathbf{E} = -\mathbf{grad}\,V.$$

Also, since $\mathrm{div}_P\,\mathbf{grad}_P\left(\dfrac{1}{r}\right) = \nabla^2\left(\dfrac{1}{r}\right) = 0$, we see that

$$\mathrm{div}\,\mathbf{D} = 0, \quad \text{or} \quad \varepsilon_0\nabla^2 V = 0. \tag{2.19, 2.20}$$

We emphasize that eqns. (2.19) and (2.20) apply only for field points which lie in regions where $\varrho = 0$, and which do not coincide with any other sort of charge, so that r cannot vanish for any point of the region of integration. Equation (2.20) is *Laplace's equation*.

Clearly, both V, $\boldsymbol{D}$ (and $\boldsymbol{E}$) are not defined by the integrals (2.14) and (2.15) for field points coinciding with point charges or surface charges, but we show that the integrals do define both V, $\boldsymbol{D}$ for field points which lie in a region where $\varrho \neq 0$ (assuming that ϱ is a bounded function of position).

A volume integral $I(n) = \iiint \dfrac{f(\xi, \eta, \zeta)\,d\tau}{r^n}$ $(n > 0)$ has an infinite integrand where $\{\xi\,\eta\,\zeta\}$, the coordinates of the volume element, coincide with $\{x\,y\,z\}$ the coordinates of the field point, if the latter lies within the volume of integra-

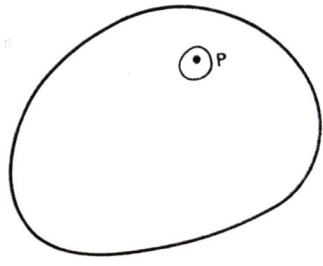

FIG. 2.10

tion. The convergence of such an integral is investigated by making a small cavity, Fig. 2.10, in which P is situated, in the volume of integration so that the function r^n does not vanish at any point of the modified volume of integration and $I(n)$ has a definite value. The integral is convergent if $I(n)$ tends to a limit as the dimensions of the cavity are made to tend to zero in an arbitrary manner. It can be shown that this limit exists if $n < 3$, i.e. the integral is convergent; there is no limit if $n > 3$, i.e. the integral is divergent. When $n = 3$ the integral $I(3)$ tends to a limit as the dimensions of the cavity are reduced to zero, but this value *depends upon the shape of the cavity used*. Hence, in this case we can define any number of "principal" values of the integral $I(3)$ by choosing special shapes for the cavity [cf. the (Cauchy) principal value of $\int_a^b f(x)\,dx$ when $f(x)$ has a singularity in the range $a < x < b$]. We apply these results to the integrals of eqns. (2.14) and (2.15).

The volume integral for V has the form $I(1)$ and is therefore convergent, and eqn. (2.15) defines V for both internal and external points. The resolutes of $\boldsymbol{D}$ (after allowing for r in the numerator) have the form of $I(2)$ and are also convergent. It follows therefore, as in (2.7), that $\boldsymbol{E} = -\mathbf{grad}\,V$. However, we cannot necessarily differentiate the integrals for $\boldsymbol{D}$ because this operation would lead to integrals of the form $I(3)$.

For future reference we note that the form of the expressions in eqns. (2.14) and (2.15) shows that, near one of the point charges Q_l,

$$V = O(1/s), \quad |\mathbf{D}| = O(1/s^2), \quad (s \to 0), \tag{2.21}$$

where s is the distance of the field point P from the charge Q_l. Also we observe that for field points at a large distance R from the origin (or from any charge, all of which we take to be at a finite distance from the origin)

$$V = O(1/R), \quad |\mathbf{D}| = O(1/R^2), \quad (R \to \infty). \tag{2.22}$$

Thus V, $|\mathbf{D}|$ together satisfy the standard boundary conditions at infinity (see *C.M.*, vol. 4, § 1.6).

2.7 Gauss's theorem

This theorem expresses Coulomb's law in a form which is applicable to any distribution of charge and not merely to the field of a single point charge. When we considered the inverse square law in § 2.4 we stated that the presence of the source charge Q_1 effectively corresponds to a total displacement across any concentric sphere equal to Q_1; the inverse square law is equivalent to spreading this displacement over a larger area as the radius of the sphere is increased, the source charge and the total displacement in the field being equal for any radius. Gauss's theorem extends this equality of source charge and total displacement to any shape of surface and any distribution of source charge. The theorem states

$$\oiint_G \mathbf{D} \cdot d\mathbf{S} = Q_G, \tag{2.23}$$

where the integral is taken over an arbitrary closed surface G (the Gauss surface) and Q_G is the total source charge enclosed by G. The integral on the left-hand side of eqn. (2.23) is the *total displacement out of G* and the theorem states that the presence of any charge corresponds to a total displacement equal to Q_G across any surface enclosing this charge.

Consider now the field produced by a number of point charges Q_l. Then

$$\mathbf{D} = \sum_l \mathbf{D}_l = \sum_l \frac{Q_l \mathbf{r}_l}{4\pi r_l^3} = -\sum_l \frac{Q_l}{4\pi} \operatorname{grad}\left(\frac{1}{r_l}\right).$$

Therefore

$$\oiint_G \mathbf{D} \cdot d\mathbf{S} = -\sum_l \frac{Q_l}{4\pi} \oiint_G \operatorname{grad}\left(\frac{1}{r_l}\right) \cdot d\mathbf{S} = -\sum_l \frac{Q_l}{4\pi} \oiint_G \frac{\partial}{\partial n}\left(\frac{1}{r_l}\right) dS.$$

§ 2.7 THE INVERSE SQUARE LAW IN ELECTROSTATICS

But the integral

$$\oint_G \frac{\partial}{\partial n}\left(\frac{1}{r_l}\right) dS = -4\pi, \quad \text{if the origin of } r_l \text{ lies inside } G;$$
$$= 0, \quad \text{if the origin of } r_l \text{ lies outside } G.$$

(See *C.M.*, vol. 4, § 1.6, p. 59.) Therefore

$$\oint_G \boldsymbol{D}\cdot d\boldsymbol{S} = \sum_G Q_l = Q_G, \tag{2.23a}$$

where $\sum_G$ denotes the sum taken over all the charges lying inside G. We extend this result to volume and surface charges by regarding each element of such distributions as a point charge; in these cases Q_G is that part of the distribution which lies inside G.

In the case of a volume charge $Q_G = \iiint_{V_G} \varrho\, d\tau$, where V_G is the volume enclosed by the surface G. Therefore

$$\oint_G \boldsymbol{D}\cdot d\boldsymbol{S} = \iiint_{V_G} \varrho\, d\tau,$$

i.e.
$$\iiint_{V_G} (\operatorname{div} \boldsymbol{D} - \varrho)\, d\tau = 0.$$

Since G is an arbitrary surface we deduce that

$$\operatorname{div} \boldsymbol{D} = \varrho. \tag{2.24}$$

This corresponds to eqn. (2.19) and applies to field points inside a volume distribution. Since $\operatorname{div} \boldsymbol{D} = \operatorname{div}(-\varepsilon_0 \operatorname{\mathbf{grad}} V)$ the equation corresponding to Laplace's equation, (2.20), is

$$\varepsilon_0 \nabla^2 V = -\varrho. \tag{2.25}$$

This is *Poisson's equation*.

In the following examples we give some important applications of Gauss's theorem.

Example 1. The field of a uniform, infinite, plane sheet of charge, of density σ.

Because the plane is of infinite extent we deduce that the field depends only on the perpendicular distance z of the field point from the charge sheet. We deduce also that at such a point the field is parallel to the normal to the plane. We make these deductions because

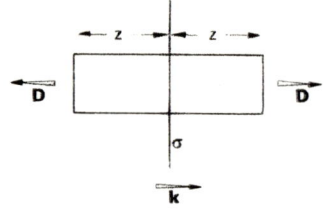

Fig. 2.11

every point of an *infinite* plane can be taken as its centre of symmetry and because of this symmetry the field at any point on the normal through the centre of symmetry must be directed along the normal and depend only on z.

We use a rectangular surface, shown in Fig. 2.11, which intercepts unit area on the plane of the charge. For this surface D is perpendicular to the normal everywhere on the side-faces and is parallel to the normal on the end-faces. Therefore

$$\oiint_G D \cdot dS = 2|D|.$$

Also $Q_G = \sigma$ the charge enclosed in G, i.e. $|D| = \tfrac{1}{2}\sigma$. Therefore

$$D_+ = k\tfrac{1}{2}\sigma, \quad D_- = -\tfrac{1}{2}k\sigma.$$

Note that D is independent of z

Example 2. The field of a uniform, infinite sheet of charge situated on a conductor.

The difference between this arrangement and that of example 1 above is that there is no field on the left-hand side of the charge sheet. Here we anticipate a result of Chapter 3, p. 71, that a non-zero electrostatic field cannot be present inside a conductor. In this case the

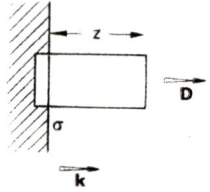

Fig. 2.12

Gauss surface is similar to that of example 1 but is closed by an end in the conductor (Fig. 2.12). The contribution to the total displacement across this surface is zero for all faces except the right-hand end-face. Therefore

$$\oiint_G D \cdot dS = |D|.$$

As before, $Q_G = \sigma$, and so $|D| = \sigma$. Therefore

$$D = k\sigma. \tag{2.26}$$

§ 2.7 THE INVERSE SQUARE LAW IN ELECTROSTATICS 43

This result is also sometimes called Coulomb's law; it is a result of considerable importance. Following from this we also have

$$E = k(\sigma/\varepsilon_0), \quad V = V_0 - z(\sigma/\varepsilon_0),$$

where V_0 is constant, the potential of the conductor.

Example 3. The field of a uniform distribution of charge of surface density σ on a long (infinite) cylinder of radius a.

Because of the cylindrical symmetry we deduce that at a distance r from the axis of the cylinder the field is a function of r only, does not vary in magnitude with the azimuthal angle, and is directed in the radial direction at right-angles to the axis of the cylinder. We choose the Gauss surface to be a concentric cylinder of unit height and radius r (Fig. 2.13).

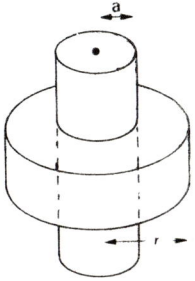

Fig. 2.13

Then the symmetry of the field implies that, on the plane faces of G, $\boldsymbol{D}$ is perpendicular to the normal and on the curved faces $\boldsymbol{D}$ is parallel to the normal, its magnitude depending only on r. Therefore

$$\oiint_G \boldsymbol{D} \cdot d\boldsymbol{S} = 2\pi r |\boldsymbol{D}|.$$

To find Q_G we consider two cases (1) $r < a$, and (ii) $r > a$.
(i) When $r < a$, $Q_G = 0$. Therefore

$$2\pi r |\boldsymbol{D}| = 0, \quad \boldsymbol{D} = \boldsymbol{0}.$$

Therefore there is no field inside a uniform, infinite, hollow cylindrical distribution of charge. The potential function is therefore a constant inside the cylinder.
(ii) When $r > a$, $Q_G = 2\pi a\sigma = e$ (the charge per unit length). Therefore

$$2\pi r |\boldsymbol{D}| = e, \quad \boldsymbol{D} = \frac{e\boldsymbol{r}}{2\pi r^2}, \quad r = |\boldsymbol{r}|,$$

since the unit radial vector is $\boldsymbol{r}/r$. We note that this result is independent of a and so applies in particular when $a = 0$. This gives an alternative, and quicker, derivation of eqn. (2.18) and all the results which follow therefrom.

Example 4. The field of a uniform, spherical surface distribution, of density σ, on a sphere of radius a (Fig. 2.14).

Because of the spherical symmetry of the distribution, we deduce that at a distance r from the centre O of the sphere the field $\boldsymbol{D}$ depends in magnitude only on r and is directed along the radius. We choose a concentric sphere of radius r as the Gauss surface. Then

$$\oint_G \boldsymbol{D} \cdot \boldsymbol{dS} = 4\pi r^2 |\boldsymbol{D}|.$$

Again we consider two cases when evaluating Q_G, viz. (i) when $r < a$, and (ii) when $r > a$.

(i) When $r < a$, $Q_G = 0$ and so $|\boldsymbol{D}| = 0$. Hence there is no field inside a spherically symmetric distribution of charge. This result applies not only to a distribution on a thin spherical shell but also to a symmetrical distribution between two concentric spheres of radii a and b. The result (i) applies because such a thick spherical arrangement can be divided into uniform shells each of infinitesimal thickness, each one contributing nothing to the field.

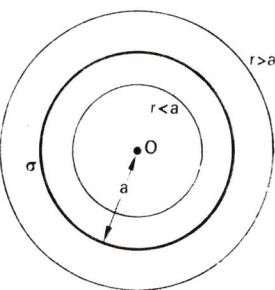

Fig. 2.14

(ii) When $r > a$, $Q_G = 4\pi a^2 \sigma = Q$ so that

$$4\pi r^2 |\boldsymbol{D}| = 4\pi a^2 \sigma = Q. \tag{1}$$

Therefore
$$\boldsymbol{D} = \frac{\sigma a^2 \boldsymbol{r}}{r^3} = \frac{Q\boldsymbol{r}}{4\pi r^3}. \tag{2}$$

This result (2) shows that, for points outside the distribution a spherically symmetrical distribution behaves as though all the charge were concentrated at the centre. As in case (i) this result applies not only to a distribution on a shell but also to a spherically symmetrical volume distribution.

Example 5. The field and potential at all points of space in the presence of a sphere of uniform volume charge density ϱ.

We apply Gauss's theorem as above using a concentric sphere of radius r as the Gauss surface. Then

$$\oint_G \boldsymbol{D} \cdot \boldsymbol{dS} = 4\pi r^2 |\boldsymbol{D}|.$$

If $r < a$,
$$Q_G = \int_0^r \varrho 4\pi s^2 \, ds = \frac{4}{3}\pi r^3 \varrho.$$

Therefore
$$\boldsymbol{D} = \tfrac{1}{3}\varrho \boldsymbol{r} = \varepsilon_0 \boldsymbol{E}. \tag{1}$$

§ 2.7 THE INVERSE SQUARE LAW IN ELECTROSTATICS

If $r > a$,
$$Q_G = \int_0^a \varrho 4\pi s^2 \, ds = \frac{4}{3}\pi a^3 \varrho = Q.$$

Therefore
$$\mathbf{D} = \frac{\varrho a^3 \mathbf{r}}{3r^3} = \frac{Q\mathbf{r}}{4\pi r^3} = \varepsilon_0 \mathbf{E}, \tag{2}$$

where Q is the total charge present in the sphere. For $r > a$ the field is identical with that of a point charge Q at the centre of the sphere. Hence,

$$V = \frac{Q}{4\pi\varepsilon_0 r} = \frac{\varrho a^3}{3\varepsilon_0 r}, \quad (r > a). \tag{3}$$

Because of the spherical symmetry we deduce, for $r < a$, that

$$-\frac{dV}{dr} = E_r = \frac{\varrho r}{3\varepsilon_0}, \quad V = V_0 - \frac{\varrho r^2}{6\varepsilon_0}.$$

But V is continuous at $r = a$ (see § 2.8), and so

$$V_0 - \frac{\varrho a^2}{6\varepsilon_0} = \frac{\varrho a^2}{3\varepsilon_0}, \quad V_0 = \frac{\varrho a^2}{2\varepsilon_0}.$$

Therefore
$$V = \frac{\varrho}{6\varepsilon_0}(3a^2 - r^2), \quad (r < a), \tag{4}$$

Equations (1) and (2) give the field vectors, and eqns. (3) and (4) give the expressions for V in the different regions. This method is easier to use, in cases where there is marked symmetry, than attempting to evaluate the integrals of eqns. (2.14) and (2.15).

Example 6. A fixed positive point charge $+e$ at a point O is surrounded by a rigid continuous distribution of negative charge whose density ϱ (<0) is a function only of the distance r from O. The total negative charge exceeds e in magnitude. Another positive point charge e' has mass m and is free to move on a line through O. Show that it can be in equilibrium at a distance r_0 from O where

$$e + 4\pi \int_0^{r_0} \varrho(r) r^2 \, dr = 0.$$

Show further that the period of small oscillations of this charge is $\sqrt{\{-\pi m \varepsilon_0 / (\varrho_0 e')\}}$, where $\varrho_0 = \varrho(r_0)$.

Using Gauss's theorem for this spherically symmetric charge distribution we deduce that, at a distance r from O,

$$4\pi r^2 D_r = e + 4\pi \int_0^r s^2 \varrho(s) \, ds.$$

Hence D_r, and E_r, vanish where $r = r_0$ provided that

$$e + 4\pi \int_0^{r_0} s^2 \varrho(s) \, ds = 0.$$

At the point $r = r_0 + \xi$,

$$4\pi(r_0 + \xi)^2 D_r = e + 4\pi \int_0^{r_0 + \xi} s^2 \varrho(s) \, ds = 4\pi r_0^2 (D_r)_{r_0} + \xi \cdot 4\pi r_0^2 \varrho(r_0)$$

correct to the first order in ξ. But $(D_r)_{r_0} = 0$, and so

$$D_r = 4\pi \xi \varrho(r_0) = 4\pi \xi \varrho_0 = \varepsilon_0 E_r,$$

correct to the first order in ξ. The equation of motion of the particle is

$$m\ddot{\xi} = e'E_r = (4\pi\varrho_0 e'/\varepsilon_0)\xi.$$

Since $\varrho_0 < 0$, this corresponds to S.H. oscillations of period

$$2\pi\sqrt{\left(-\frac{m\varepsilon_0}{4\pi\varrho_0 e'}\right)} = \sqrt{\left(-\frac{\pi m\varepsilon_0}{\varrho_0 e'}\right)}.$$

Tubes of Force

When a field is established between a number of conductors which carry charges, there being no volume charge, the lines of force arise from positive charge and end on negative charge. We construct a tube-like surface by draw-

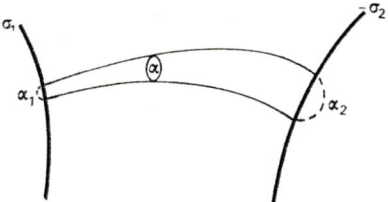

Fig. 2.15

ing the lines of force through the perimeter of a small area α in the field (Fig. 2.15). When the lines are continued to meet the conductors at each end, the tube intersects an area α_1 on the conductor where the (positive) charge density is σ_1, and α_2 on the conductor where the (negative) charge density is $-\sigma_2$. We can close the surface by drawing caps in the material of the conductor at each end. The total displacement out of this closed (Gauss) surface is zero; the contribution is zero from the ends where $\boldsymbol{D} = \boldsymbol{0}$ in the conductors; the contribution is zero from the remaining surface where D lies in the surface. Hence, by Gauss's theorem,

$$Q_G = 0 = \alpha_1\sigma_1 - \alpha_2\sigma_2, \quad \text{i.e.} \quad \alpha_1\sigma_1 = \alpha_2\sigma_2,$$

so that the charges on which a tube begins and ends are equal and opposite. We take a *unit tube* to be one arising from unit charge; in general, the *strength* of a tube is the positive charge from which it arises. If we close the tube by a cap α_1 and by a normal section α where the displacement is $\boldsymbol{D}$, then for this Gauss surface

$$\oint_G \boldsymbol{D}\cdot d\boldsymbol{S} = \alpha|\boldsymbol{D}|, \quad Q_G = \alpha_1\sigma_1 = N,$$

§ 2.7 THE INVERSE SQUARE LAW IN ELECTROSTATICS

where N is the number of unit tubes starting from α_1 and passing through α. Therefore

$$|D| = N/\alpha.$$

Hence we may interpret $|D|$ as the density of unit tubes. This concept of a *unit tube* makes more definite the qualitative idea of a line of force; the convergence of lines of force on a negative charge corresponds to an increasing density of unit tubes nearer to the charge and so corresponds to increasing field strength nearer to the charge. By this means we can form a picture of an electrostatic field in which each positive charge has the appropriate number of unit tubes diverging from it; these same tubes must all end on equal negative charge somewhere else in the field. In this context Gauss's theorem states that if a surface G encloses a net charge Q_G, then Q_G unit tubes must cross the surface in order to end on negative charges, amounting to $-Q_G$, outside G.

With this (more precise) idea of tubes of force we give a more detailed description of the field produced by some special distributions of charge.

Example 1. A field is produced by a series of collinear point charges Q_l at the points A_l, ($l = 1, 2, \ldots, M$). Then, with the notation of Fig. 2.16, the equations of the lines of force are

$$\sum_1^M Q_l \cos \theta_l = \text{constant}.$$

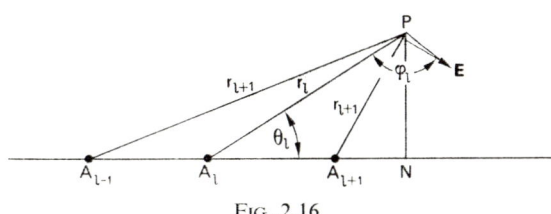

Fig. 2.16

Instead of using the differential equations given earlier for the lines of force, we resolve the fields produced by the individual charges in a direction perpendicular to E (or D) at an arbitrary field point P. Since the system has cylindrical symmetry, the result obtained applies to any plane containing the axis $A_1 A_2 \ldots A_M$. The resolution perpendicular to E gives

$$\sum_1^M \frac{Q_l \sin \phi_l}{4\pi r_l^2} = 0.$$

But
$$PN = r_1 \sin \theta_1 = r_2 \sin \theta_2 = \ldots = r_M \sin \theta_M, \tag{1}$$

and
$$\sin \phi_1 = r_1 \frac{d\theta_1}{ds}, \quad \sin \phi_2 = r_2 \frac{d\theta_2}{ds}, \ldots, \sin \phi_M = r_M \frac{d\theta_M}{ds}, \tag{2}$$

where the differentiation d/ds is taken along the line of force passing through P. Therefore

$$\sum_1^M \frac{Q_l}{4\pi r_l} \frac{d\theta_l}{ds} = 0.$$

Eliminating r_l by the use of eqn. (1) we find that

$$\sum_1^M \frac{Q_l}{4\pi} \sin\theta_l \cdot \frac{d\theta_l}{ds} = 0,$$

which we can integrate w.r. to s, to give

$$\sum_1^M Q_l \cos\theta_l = \text{constant}.$$

Different values of the constant give different lines of force.

Example 2. As a special case of the above we consider the case of three collinear charges $q, -2q, q$ at A, B, C respectively (see Figs. 2.16 and 2.17). Show that a line leaving A at an angle ϕ with AB ends on B at an angle θ with BA where

$$\sin \tfrac{1}{2}\phi = \sqrt{2} \sin \tfrac{1}{2}\theta.$$

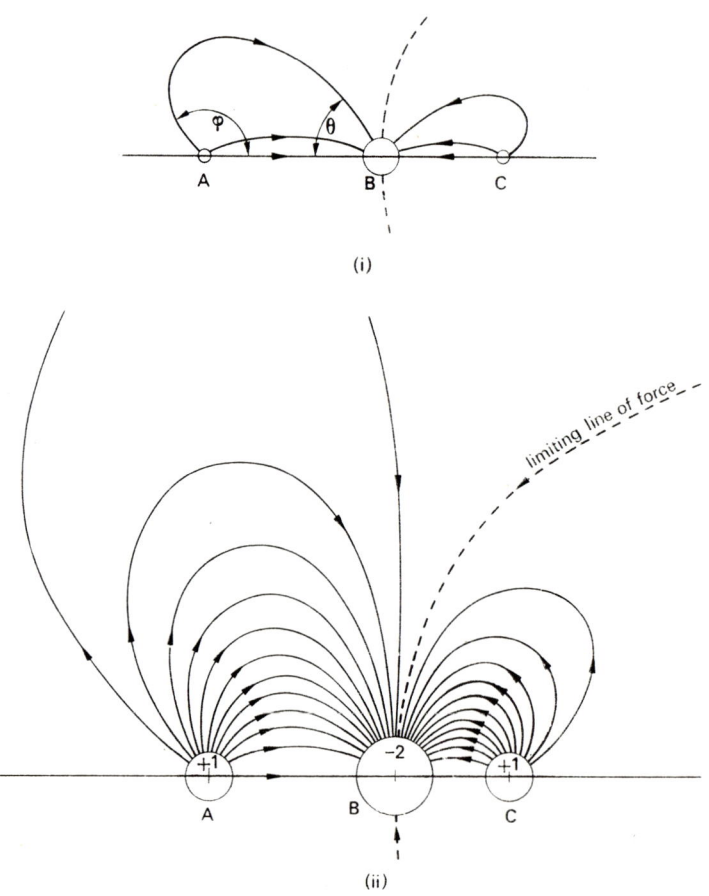

Fig. 2.17

(The distribution of the lines in Figs. 2.16 and 2.17 is slightly misleading because a *2-dimensional* representation of a distribution which is uniform in *3 dimensions* near the point charge does not appear uniform in the diagram.)

Because we are now attributing the quantitative significance of unit tubes to figures generated by lines of force we see that we can divide the tubes ending on B into two groups, those starting from A and those from C. (Note that $AB \neq BC$ in general.) These tubes are separated by a dividing surface whose section is shown dotted in Fig. 2.17. (If $AB = BC$ the figure must be symmetrical about B and the dividing surface is the plane of symmetry through B.) Since the tubes of force very close to B represent the field of a point charge, that of B, which dominates the effects of the other charges, the tubes enter B uniformly distributed in space around B. Since half the tubes originate on each positive charge the dividing surface has its tangent plane at B perpendicular to ABC.

By the previous example the equation of a line of force is

$$\cos\theta_1 - 2\cos\theta_2 + \cos\theta_3 = \text{constant} = C_1.$$

(We use suffixes 1, 2, 3 for the charges at A, B, C respectively.) Near A, for the line leaving at the angle ϕ,

$$\theta_1 = \phi, \quad \theta_2 = \pi, \quad \theta_3 = \pi,$$

and for this line of force $C_1 = \cos\phi + 2 - 1 = 1 + \cos\phi$. Near B, for the same line, $\theta_1 = 0$, $\theta_2 = \pi - \theta$, $\theta_3 = \pi$. Therefore

$$1 + 2\cos\theta - 1 = 1 + \cos\phi.$$

This gives the required result.

Example 3. We consider the case of two unequal charges $4q$, $-q$ separated by a distance a (see Figs. 2.16 and 2.18).

Since more unit tubes leave A than can enter B, only one-quarter of those leaving A end on B, the remainder end at infinity. Hence there is a surface dividing these two sets of tubes. This surface must subtend a solid angle $\frac{1}{4} \cdot 4\pi$ at A symmetrically disposed round AB. On the section shown in Fig. 2.18, this corresponds to an angle α, where

$$2\pi(1-\cos\alpha) = \pi, \quad \cos\alpha = \tfrac{1}{2}, \quad \alpha = \tfrac{1}{3}\pi.$$

The equation of a line of force is

$$4\cos\theta_1 - \cos\theta_2 = \text{constant} = C, \tag{1}$$

and for a line on this dividing surface at A, $\theta_1 = \alpha$, $\theta_2 = \pi$; and so for this line $C = 4\cos\alpha - \cos\pi = 3$, i.e. the critical line has equation

$$4\cos\theta_1 - \cos\theta_2 = 3.$$

Now this equation is also satisfied by $\theta_1 = \theta_2 = 0$, so that the line AB extended beyond B is part of this same critical line of force. In this field there is a neutral point N ($BN = a$) and the direction of the field on either side of N is as shown. We conclude that the critical line passes through N, together with a symmetrical branch below AB, and divides at N. (An evaluation of the field close to N shows that the "cross" is made at right angles.)

In the mapping of fields in this manner the determination of the neutral point and the critical line of force is important. When these are known the general shape of the field is then reasonably clear.

Note. From eqn. (1) we see that the equation of a line of force leaving A at an angle β with AB is

$$4\cos\theta_1 - \cos\theta_2 = 1 + 4\cos\beta.$$

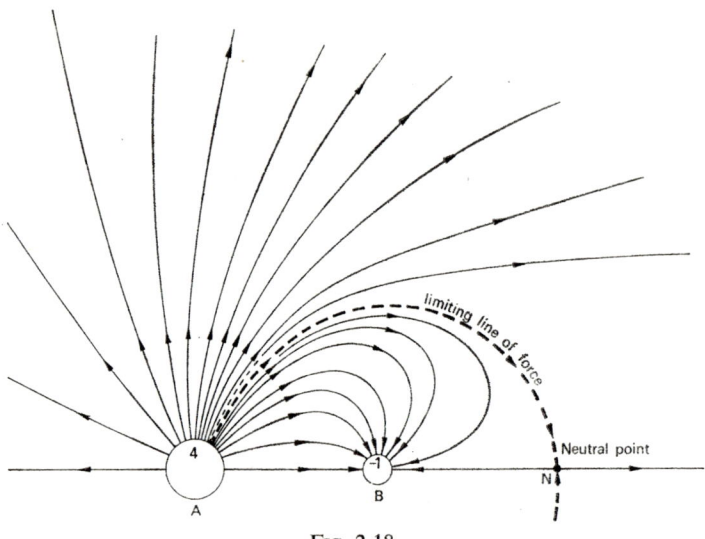

Fig. 2.18

If this line goes to infinity it does so in a direction γ where $\theta_1 = \theta_2 = \gamma$ and

$$3 \cos \gamma = 1 + 4 \cos \beta.$$

This equation only has a real solution for γ if, and only if,

$$-1 \leqslant \cos \beta \leqslant \tfrac{1}{2}.$$

Here γ gives the asymptotic direction of a line of force.

Exercises 2.7

1. Prove that the equation of a line of force in a two-dimensional field produced by a number of line charges $e_1, e_2, e_3, \ldots, e_N$, at $A_1, A_2, A_3, \ldots, A_N$ is

$$\sum_1^N e_N \theta_N = \text{constant}$$

where the notation is the same as that in the example of p. 37.
   Show further that the equation is true, for a two-dimensional field, even when $A_1, A_2, \ldots A_N$ are not collinear.

2. Use the result of question 1 to show that the lines of force joining two line charges $\pm e$ are arcs of circles which intersect on the charges. What are the shapes of the equipotential surfaces?

3. Show that the potential due to a radially symmetric charge distribution in space of density

$$\varrho = \frac{\varrho_0 a}{\pi r} \sin\left(\frac{\pi r}{a}\right), \quad 0 \leqslant r \leqslant a,$$
$$\varrho = 0, \quad r \geqslant a,$$

is given by

$$4\pi\varepsilon_0 V = \frac{Q}{a}\left(1+\frac{\varrho}{\varrho_0}\right), \qquad 0 \leqslant r \leqslant a,$$

$$4\pi\varepsilon_0 V = \frac{Q}{r}, \qquad r \geqslant a,$$

where Q is the total charge.

4. From a sphere, of uniform charge density ϱ, radius a and centre O, is removed a sphere of radius b and centre O_1, where $OO_1 = c\,(< a-b)$. Use the result of example 5, p. 44, to show that the field at any point P inside the cavity is $\mathbf{D} = \tfrac{1}{3}\varrho\,(\overrightarrow{OP}-\overrightarrow{O_1P})$. Deduce that within the cavity $\mathbf{D}$ is uniform, parallel to $\overrightarrow{OO_1}$, and of magnitude $\tfrac{1}{3}\varrho c$.

5. A number of point charges $e_1, e_2, e_3, \ldots$, are fixed at interior points $Q_1, Q_2, Q_3, \ldots$ of a line OX. Show that if P is a point on any selected line of force, then

$$\sum e_r \cos \theta_r = \text{constant},$$

where $\theta_r = PQ_rX$.

Electric charge is distributed uniformly along a line AB of finite length. Show that if P is a point on a line of force, then

$$PA - PB = \text{constant}.$$

Hence show that in any plane through AB the equipotential curves are ellipses with A and B as foci.

6. Charges q and $-\lambda q$ $(0 < \lambda < 1)$ are placed on the y-axis at the points $y = -a$ and $y = a$, respectively. Sketch the lines of force and prove that all the lines of force from the charge $-\lambda q$ cross the plane $y = 0$ at a distance from the origin not greater than $2a(\sqrt{\lambda})/(1-\lambda)$.

7. Positive charges q_1 and q_2 are placed at points A and B respectively. A line of force starts from A at an angle α to BA produced. Prove that its asymptote passes through the point C on AB such that $AC/CB = q_2/q_1$, and makes an angle β with BA given by

$$\sin\left(\frac{\beta}{2}\right) = \left(\frac{q_1}{q_1+q_2}\right)^{1/2} \sin\left(\frac{\alpha}{2}\right).$$

2.8 Discontinuities and boundary conditions

The sources of a field occur not only as singularities (point charges) but also as surface and volume charges. In § 2.6 we excluded consideration of field points on such surfaces. Discontinuities may also occur at the boundary of a region in which $\varrho \neq 0$, for across such a boundary ϱ changes discontinuously. In this section we investigate how such discontinuities affect the field quantities $\mathbf{D}, \mathbf{E}, V$. We cannot usually give values to these quantities at points on the surfaces of discontinuity, or at points which coincide with point charges, but we can consider the values of the field quantities at points indefinitely close to and on either side of a surface of discontinuity. Our usual notation is indicated in Fig. 2.19 and has been anticipated in one or two examples earlier (see pp. 35 and 42). The diagram shows a section of the surface of discontinuity, the unit normal $\hat{\boldsymbol{\nu}}$ distinguishing a positive and a negative side of the surface. We consider two field points P_+, P_- on the respective sides of the surface, on the same normal, such that their distances from the surface are quantities of the *second* order. This means that when we mark off a small area α of the surface, whose dimensions are of the *first* order $O(l)$, then for the field points P_+ and P_- this area (indicated by double thickness on Fig. 2.19) may be considered as an infinite plane area, i.e. $P_+ P_- = O(l^2)$.

The expressions in eqns. (2.14) and (2.15) give $\mathbf{D}, \mathbf{E}, V$ as functions of the position of the field point P; when P is taken to coincide with P_+, or with P_-, in Fig. 2.20 any discontinuity in one of these field quantities will be shown as a (finite) difference in the values at these two points. The general technique used in our investigations resembles that used in § 2.6. We divide the region of integration into two parts, one small part close to P_+ and P_- and the "remainder". As in § 2.6 the contribution from the "remainder"

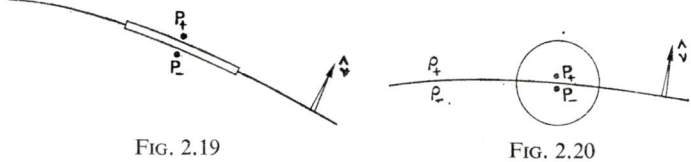

Fig. 2.19   Fig. 2.20

will be identical at P_+ and P_-, correct to the first order, since both P_+ and P_- are points external to the region of integration and are separated by a distance of the second order. Hence in calculating discontinuities we need to consider only the "local" contribution.

When there is a discontinuity in the volume density ϱ any discontinuity in the field quantities arises solely from the volume integrals in eqns. (2.14), (2.15). We remove a small region from each side of the surface, so as to enclose the points P_+ and P_-, as shown in Fig. 2.20. We assume that ϱ is bounded on each side of the surface. In order to investigate the discontinuities in V and $\mathbf{D}\,(=\varepsilon_0 \mathbf{E})$ we must evaluate integrals such as

$$\iiint \frac{\varrho_+ \,\mathrm{d}\tau}{r}, \quad \iiint \frac{\varrho_+ \,\mathrm{d}\tau}{r^2}, \quad \iiint \frac{\varrho_- \,\mathrm{d}\tau}{r}, \quad \iiint \frac{\varrho_- \,\mathrm{d}\tau}{r^2}$$

for points P_+ and P_-, the integrals being taken through the small volume which includes these points. Then we must find the limiting values of these integrals as the volume tends to zero. Since ϱ_+ and ϱ_- are both bounded, and all the integrals are convergent [being of the form $I(1)$ or $I(2)$] they must all tend to zero as the volume of integration tends to zero. We conclude that *all* the field quantities are continuous across a discontinuity in ϱ.

We investigate the discontinuities produced by a surface distribution of charge by dividing the surface into a small area near P_+ and P_- and the "remainder" (see Fig. 2.19). We suppose that the displacement due to the

§ 2.8 THE INVERSE SQUARE LAW IN ELECTROSTATICS

"remainder" is D' and we quote the result of example 3, p. 35 (or example 1, p. 41) for the "local" contribution. As pointed out above, for the field points P_+ and P_- the local contribution is effectively that of an infinite plane. Hence

$$D_+ = D' + \tfrac{1}{2}\sigma\hat{v}, \quad D_- = D' - \tfrac{1}{2}\sigma\hat{v}.$$

Therefore
$$D_+ - D_- = \sigma\hat{v}. \tag{2.27}$$

This leads to the following results

$$\hat{v} \cdot (D_+ - D_-) = \sigma \tag{2.28}$$

and, in terms of E,

$$\hat{v} \times (E_+ - E_-) = 0. \tag{2.29}$$

The forms (2.28) and (2.29) are the forms usually quoted, and they can be obtained separately as shown below.

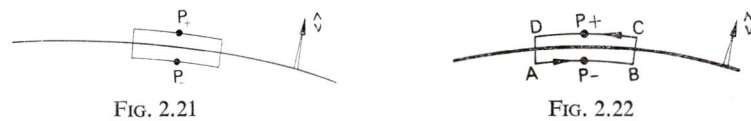

FIG. 2.21   FIG. 2.22

Equation (2.28) is the surface form of the relation div $D = \varrho$; to show that this is so we apply Gauss's theorem to a "pill-box" shaped surface whose plane faces pass through P_+ and P_- and whose sides are generated by normals to the surface (Fig. 2.21) drawn through the perimeter of a small area α. Since the perimeter of the area is of the first order, the area generated by the normals is of the third order and the contribution to the total displacement $\oint_G D \cdot dS$ from this side is of the third order. Hence

$$\oint_G D \cdot dS = (D_+ - D_-) \cdot \hat{v}\alpha + O(l^3).$$

But for this Gauss surface $Q_G = \alpha\sigma$. Therefore

$$\hat{v} \cdot (D_+ - D_-) = \sigma, \tag{2.28}$$

The expression $\hat{v} \cdot (D_+ - D_-)$ is called the *surface divergence*, denoted by Div D, so that eqn. (2.28) is the surface counterpart of eqn. (2.24). [See also C.M., vol. 4, § 1.7(1).] (This derivation is to be preferred to that from eqn. (2.27) for that result depended upon using a circular area with P_+P_- as its axis. The proof using Gauss's theorem does not require this to be so.)

The result in eqn. (2.29) is similarly the surface counterpart of eqn. (2.6). We consider a small closed contour $ABCD$ passing through P_+ and P_- (Fig. 2.22) such that AB, CD are of the first order and BC, DA of the second order.

We calculate the value of $\oint \mathbf{E} \cdot d\mathbf{S}$ for this contour; since the ends are of negligible length compared with the sides and $|\mathbf{E}|$ is bounded, we can neglect the contribution to the line integral from BC, DA. If we denote each of the lengths DC, AB by l and if the unit vector $\hat{e}$ perpendicular to the area $ABCD$ is connected with the circulation by the right-hand rule, then

$$\overrightarrow{AB} = l(\hat{v} \times \hat{e}), \quad \overrightarrow{CD} = -l(\hat{v} \times \hat{e}).$$

$$\oint \mathbf{E} \cdot d\mathbf{S} = l(\hat{v} \times \hat{e}) \cdot (-\mathbf{E}_+ + \mathbf{E}_-) + O(l^2).$$

But $\oint \mathbf{E} \cdot d\mathbf{S} = 0$, by the conservation of energy, for an arbitrary contour. Therefore $\hat{e}$ may be chosen arbitrarily and we deduce that

$$\hat{v} \times (\mathbf{E}_+ - \mathbf{E}_-) = \mathbf{0}. \tag{2.29}$$

The expression on the left-hand side is the *surface curl*, **Curl E**, and eqn. (2.29) is the surface counterpart of eqn. (2.6).

Since the charge on the element of area α in Fig. 2.19 is situated in the field of the "remainder" of the charge, we expect it to experience a force given by

$$\mathbf{F} = (\mathbf{D}'/\varepsilon_0)\sigma\alpha = \frac{1}{2\varepsilon_0}(\mathbf{D}_+ + \mathbf{D}_-)\sigma\alpha = \frac{1}{2}(\mathbf{E}_+ + \mathbf{E}_-)\sigma\alpha. \tag{2.30}$$

In the special case of a charge on a conductor, see example 2, p. 42, $\mathbf{E}_+ = \hat{v}\sigma/\varepsilon_0$, $\mathbf{E}_- = \mathbf{0}$. Hence the force on an area α of the conductor is

$$\mathbf{F} = \frac{\alpha \sigma^2}{2\varepsilon_0} \hat{v}. \tag{2.31}$$

This is a normal, outward pressure $\tfrac{1}{2}\sigma^2/\varepsilon_0$.

The effect on the expression for the potential of the local area of Fig. 2.19 shows that

$$V_+ = V' + O(l^2), \quad V_- = V' + O(l^2)$$

where α ($\approx l^2$) is the local area. Hence in the limit

$$V_+ - V_- = 0. \tag{2.32}$$

We can also express eqn. (2.28) in terms of V in the form

$$\left(\frac{\partial V}{\partial n}\right)_+ - \left(\frac{\partial V}{\partial n}\right)_- = -\frac{\sigma}{\varepsilon_0}. \tag{2.33}$$

This is the surface form of Poisson's equation and eqns. (2.32) and (2.33) are together equivalent to eqns. (2.28) and (2.29).

2.9 Applications of Green's theorem

The general form of the inverse square law, Gauss's theorem, is the basis of the results, including Poisson's equation, which refer to continuous volume and surface distributions of charge. The derivation given in § 2.7 involved the following steps:

$$\oint_G \mathbf{D} \cdot \mathbf{dS} = \oint_G (\sum_l \mathbf{D}_l) \cdot \mathbf{dS} = \sum_l \oint_G \mathbf{D}_l \cdot \mathbf{dS} = \sum_G Q_l = Q_G.$$

The final result was obtained by interchanging the operations $\oint\!\!\!\int \ldots$ and $\sum_l$, shown in the middle two terms above. When we make the transition from a finite number of point charges to a continuous distribution as the source of a

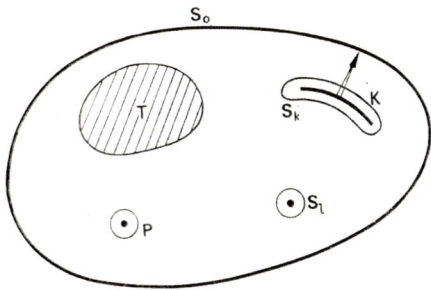

Fig. 2.23

field, as in the derivation of eqn. (2.24) or subsequent applications, the operation $\sum_l$ becomes itself a limiting process, effectively an integration. Hence application to a continuous distribution of that form of Gauss's theorem, proved in eqn. (2.23a), is open to objection on pure mathematical grounds, viz. the commutation of two limiting processes without an investigation of whether this interchange is permissible. We now justify this step by the use of Green's theorem.

Consider a distribution of charge in the form of volume, surface, or point charges, the magnitude of the total charge being finite. The charge is all at a finite distance from the origin, so that we can enclose all the charge in a surface S_0 by choosing S_0 sufficiently large. In Fig. 2.23 specimens of each kind of charge are indicated; volume charge is in the volume T where $\varrho \neq 0$; surface charge σ is on a surface labelled K which is enveloped by a closed surface S_k, and a point charge is enclosed by a surface S_l. The field point P is enclosed by a small sphere C (not labelled in Fig. 2.23). We apply Green's theorem in the form

$$\iiint \frac{\nabla^2 V}{r} d\tau = \oiint \left\{ \frac{1}{r} \frac{\partial V}{\partial n} - V \frac{\partial}{\partial n}\left(\frac{1}{r}\right) \right\} dS$$

to V, the potential function of the field, using the region enclosed by S_0 but lying outside S_k, S_l, and C. The volume of integration includes T but excludes the surface and point charges and the point P. The vector $\mathbf{r}$ is drawn from the element of integration to P, $r = |\mathbf{r}|$, and the direction denoted by the differentiation $\partial/\partial n$ is *out* of the region of integration.

We assume that both V and its derivatives are bounded at P, and that they have the properties given by eqns. (2.21) and (2.22). We note the following facts concerning the integrals in the equation above:

(1) Since $\nabla^2 V = 0$ at all points of the region where $\varrho = 0$ the left-hand side becomes

$$\iiint_{T_d} \frac{\nabla^2 V}{r} d\tau.$$

(2) Because V satisfies the standard boundary conditions at infinity, the contribution to the right-hand side from S_0 tends to zero as $R \to \infty$, i.e. as S_0 is moved to infinity [eqn. (2.22)].

(3) From the properties given in eqn. (2.21) we see that, since $1/r$ and $\partial(1/r)/\partial n$ are finite over S_l,

$$\oiint_{S_l} V \frac{\partial}{\partial n}\left(\frac{1}{r}\right) dS \to 0; \quad \oiint_{S_l} \frac{1}{r} \frac{\partial V}{\partial n} dS \approx \frac{1}{r_l} \oiint_{S_l} \frac{\partial V}{\partial n} dS \to \text{a finite limit}$$

as S_l is reduced to zero dimensions on the point charge.

(4) The function $1/r$ is continuous across the surface K and $\dfrac{\partial}{\partial n}\left(\dfrac{1}{r}\right)$ has equal and opposite values on the two sides of the envelope S_k. Hence

$$\oiint_{S_k} \frac{1}{r} \frac{\partial V}{\partial n} dS \to \iint_K \frac{1}{r} \left\{ \left(\frac{\partial V}{\partial n}\right)_- - \left(\frac{\partial V}{\partial n}\right)_+ \right\} dS; \quad \oiint_{S_k} V \frac{\partial}{\partial n}\left(\frac{1}{r}\right) dS \to 0,$$

as the surface S_k tends to coincidence with the surface K. We assume in the second of these results that V is continuous across the surface K. (See, however, question 2 in Exercises 2.12.)

(5) Since V, $\partial V/\partial n$ are finite at P,

$$\oiint_C \frac{1}{r} \frac{\partial V}{\partial n} dS = -\oiint_C \frac{1}{r} \frac{\partial V}{\partial r} dS, \quad \left(\frac{\partial}{\partial n} \equiv -\frac{\partial}{\partial r} \quad \text{on} \quad C\right).$$

§ 2.9 THE INVERSE SQUARE LAW IN ELECTROSTATICS

Therefore
$$\left| \oint_C \frac{1}{r} \frac{\partial V}{\partial n} dS \right| \leq m \oint \frac{r^2 d\omega}{r} \to 0, \text{ as } r \to 0,$$

where m is the maximum value of $|\partial V/\partial r|$ on C. Also

$$\oint_C V \frac{\partial}{\partial n}\left(\frac{1}{r}\right) dS = \oint_C \frac{V}{r^2} r^2 d\omega \to 4\pi V(P)$$

as the radius of C tends to zero. (The element $d\omega$ is the solid angle subtended by the element of area dS on C, i.e. $dS = r^2 d\omega$.)

Collecting these results (1)–(5) and taking the limiting case as $S_0 \to \infty$, $S_k \to K$, $S_l \to 0$, $C \to 0$, we obtain

$$\iiint_T \frac{\nabla^2 V}{r} d\tau = \frac{1}{r_l} \lim \oint_{S_l} \frac{\partial V}{\partial n} dS - \iint_K \frac{1}{r}\left\{\left(\frac{\partial V}{\partial n}\right)_+ - \left(\frac{\partial V}{\partial n}\right)_-\right\} dS - 4\pi V(P).$$

Therefore
$$V(P) = -\iiint_T \frac{\nabla^2 V}{4\pi r} d\tau - \iint_K \frac{1}{4\pi r}\left\{\left(\frac{\partial V}{\partial n}\right)_+ - \left(\frac{\partial V}{\partial n}\right)_-\right\} dS$$
$$+ \frac{1}{4\pi r_l} \lim \oint_{S_l} \frac{\partial V}{\partial n} dS.$$

But eqn. (2.15) for V gives

$$V(P) = \iiint_T \frac{\varrho \, d\tau}{4\pi\varepsilon_0 r} + \iint_K \frac{\sigma \, dS}{4\pi\varepsilon_0 r} + \frac{Q_l}{4\pi\varepsilon_0 r}.$$

A comparison of the last two equations shows that

$$\nabla^2 V = -\frac{\varrho}{\varepsilon_0}, \quad \left(\frac{\partial V}{\partial n}\right)_+ - \left(\frac{\partial V}{\partial n}\right)_- = -\frac{\sigma}{\varepsilon_0}, \quad \lim \iint_{S_l} \frac{\partial V}{\partial n} dS = \frac{Q_l}{\varepsilon_0}. \quad (2.34)$$

The first of eqns. (2.34) is Poisson's equation, from which we can deduce, by integration throughout the volume T_G bounded by an arbitrary Gauss surface G,

$$\iiint_{T_G} \varrho \, d\tau = -\varepsilon_0 \iiint_{T_G} \nabla^2 V \, d\tau$$
$$= \iiint_{T_G} \operatorname{div}(-\varepsilon_0 \operatorname{\mathbf{grad}} V) \, d\tau = \iiint_{T_G} \operatorname{div} \mathbf{D} \, d\tau.$$

Therefore
$$\operatorname{div} \mathbf{D} = \varrho.$$

By applying the divergence theorem to the volume integral we obtain

$$\iiint_{T_G} \varrho \, d\tau = \oiint_G \boldsymbol{D} \cdot d\boldsymbol{S},$$

which shows that Gauss's theorem holds for volume distributions of charge.

The second of eqns. (2.34) is the surface form of Poisson's equation, which can be written

$$\hat{\boldsymbol{\nu}} \cdot (\boldsymbol{D}_+ - \boldsymbol{D}_-) = \sigma$$

and so confirms that Gauss's theorem applies also to surface distribution of charge.

The third of eqns. (2.34) is equivalent to eqn. (2.21) and is effectively Gauss's theorem for point charges, a result already proved adequately.

2.10 Dipoles and multipoles

When a positive charge Q is separated from a charge $-Q$ by a small displacement $\boldsymbol{l}$, for field points at large distances the two charges effectively coincide in position, since the distance $l = |\boldsymbol{l}|$ is negligible compared with

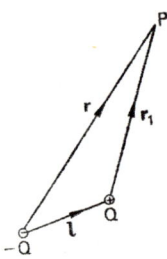

Fig. 2.24

the distance r to the field point (Fig. 2.24). Although the charges effectively coincide in position, their electrostatic effects do not cancel. In order to study the effects of such a pair of charges we consider the limiting case for which $Q \to \infty$, $l \to 0$ in such a way that μ is finite where

$$\mu = \lim Q\boldsymbol{l}. \tag{2.35}$$

Such an arrangement is called an electric *dipole* with *moment* μ.

We calculate the field of which a dipole is the sole source by applying the limiting process (2.35) to the potential V, and to the fields $\boldsymbol{D}, \boldsymbol{E}$ produced by

§ 2.10   THE INVERSE SQUARE LAW IN ELECTROSTATICS   59

the point charges of Fig. 2.24. From eqns. (2.12) and (2.13),

$$D = \lim \frac{Q}{4\pi}\left(\frac{r_1}{r_1^3} - \frac{r}{r^3}\right) = \varepsilon_0 E, \quad V = \lim \frac{Q}{4\pi\varepsilon_0}\left(\frac{1}{r_1} - \frac{1}{r}\right).$$

The scalar $1/r_1 - 1/r$ is the change in the scalar $1/r$ corresponding to the displacement l. Therefore,

$$\frac{1}{r_1} - \frac{1}{r} = l \cdot \mathrm{grad}_S\left(\frac{1}{r}\right),$$

the differentiation being carried out w.r. to the coordinates of the source point S. Therefore

$$V = \lim \frac{Ql}{4\pi\varepsilon_0} \cdot \mathrm{grad}_S\left(\frac{1}{r}\right) = -\lim \frac{Ql}{4\pi\varepsilon_0} \cdot \mathrm{grad}_P\left(\frac{1}{r}\right) = \frac{\mu \cdot r}{4\pi\varepsilon_0 r^3}. \quad (2.36)$$

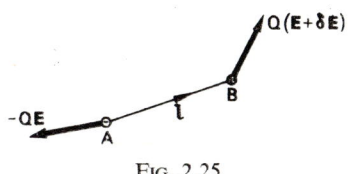

Fig. 2.25

Here we have used the result that $\nabla_S = -\nabla_P$. (See *C.M.*, vol. 4, p. 67.) Since $r/r^3 = -\mathrm{grad}_P(1/r)$ we see that

$$\varepsilon_0 E = D = -\varepsilon_0 \mathrm{grad}\, V = -\frac{\mu}{4\pi r^3} + \frac{3(\mu \cdot r)r}{4\pi r^5}. \quad (2.37)$$

When a dipole is placed in a field arising from other sources, the charge at each end of the dipole experiences a force given by the electrostatic field prevailing at that point. The action of the field on a dipole therefore consists of two forces, in general skew to each other. We reduce these two forces to a resultant F at A and a couple $\Gamma(A)$ about A. In the position shown in Fig. 2.25 the force and couple are

$$Q(-E+E+\delta E), \quad Ql \times (E+\delta E).$$

But
$$\delta E = (l \cdot \nabla)E;$$
therefore
$$F = \lim Q(l \cdot \nabla)E = (\mu \cdot \nabla)E, \quad (2.38)$$
$$\Gamma(A) = \lim Ql \times (E+\delta E) = \mu \times E. \quad (2.39)$$

A dipole, in the strict limiting sense as defined above, cannot occur in practice, but it is nevertheless a useful concept. Consider a charge Q at a point A (Fig. 2.26) not coinciding with the origin, where $\overrightarrow{OA} = l$. For a field point

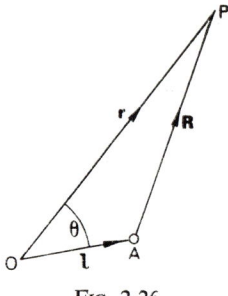

Fig. 2.26

P at a distance from O and A large compared with OA,

$$V = \frac{Q}{4\pi\varepsilon_0 R} = \frac{Q}{4\pi\varepsilon_0 \sqrt{(r^2 - 2rl\cos\theta + l^2)}}.$$

Since l/r is small we can expand this expression by the binomial theorem, or by the methods of *C.M.*, vol. 4, p. 202, to give

$$V = \frac{Q}{4\pi\varepsilon_0}\left\{\frac{1}{r} + \frac{l\cos\theta}{r^2} + O\left(\frac{l^2}{r^3}\right)\right\}.$$

We can therefore obtain successive approximations, $V_1, V_2, \ldots,$ to the potential function, the first two being

$$V_1 = \frac{Q}{4\pi\varepsilon_0 r}, \quad V_2 = V_1 + \frac{Ql\cos\theta}{4\pi\varepsilon_0 r^2}.$$

The potential V_1 is that of a point charge Q at O, instead of at A. The first correction to this is the term

$$\frac{Ql\cos\theta}{4\pi\varepsilon_0 r^2} = \frac{Q(\mathbf{l}\cdot\mathbf{r})}{4\pi\varepsilon_0 r^3}.$$

This is the potential of a dipole of moment $Q\mathbf{l}$ at O and shows that the second approximation regards the charge Q at A as equivalent to Q at O and also $-Q$ at O with Q at A taken together as a dipole. (The process resembles the reduction of a set of forces by the addition of null pairs at the base point.) Higher order corrections are given by higher order multipoles.

The next higher order is a quadrupole; it can be regarded as the limit of two dipoles. The two dipoles, separated by $\mathbf{l}_1$, are made to coincide and the magnitudes $\mu = |\boldsymbol{\mu}| \to \infty$ and $l_1 = |\mathbf{l}_1| \to 0$ so that the product $\mu l_1 = \mu_1$, where μ_1 is finite. As indicated in Fig. 2.27 the moment of a quadrupole contains two directions (which may coincide); it has to be represented by a *dyadic* or tensor of order 2.

When a charge distribution is situated in the neighbourhood of a point O, for distant field points it may be regarded as a point charge at O together with a resultant dipole, a resultant quadrupole, etc., according to the degree of approximation, all situated at O. When the total charge in the distribution is zero then the dipole moment predominates; when both total charge and dipole moment are zero the quadrupole moment predominates. The arrangement of three charges (when equally spaced apart) in example 2, p. 48, is

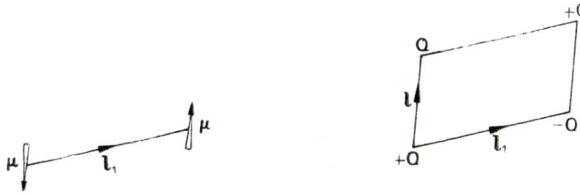

FIG. 2.27

one which at large distances produces a field corresponding to a quadrupole moment. Since an atom consists of a positive nucleus, surrounded by a neutralizing distribution of electrons, it may have a dipole moment; if the electrons are symmetrically distributed, the dipole moment is zero, leaving a quadrupole moment as the sole electric effect. This analysis into multipoles giving the effect at large distances is an important reason for investigating what, at first sight, seems a highly artificial combination of charges.

2.11 Double layers

In § 2.4, p. 32, we considered (simple) charge distributed as point, surface, or volume charges. Here we consider dipoles distributed similarly. A surface distribution of dipoles, called a *double layer*, is obtained by the coincidence of two layers of surface charge of strengths $\pm\sigma$, Fig. 2.28. We consider one type of layer only, that produced by two simple layers so arranged that the resultant dipole moment at a typical point R is directed along the normal $\hat{\nu}$ to the surface. This is equivalent to making the two simple layers approach along the normal and letting $\sigma \to \infty$ and the separation $t \to 0$ so that

$$\tilde{\omega} = \lim \sigma t \tag{2.40}$$

is finite. The dipole moment of a small area α of the resultant double layer is then $\tilde{\omega}\hat{\nu}\alpha$, correct to the first order. The double layer is *uniform* if $\tilde{\omega}$ is constant over the surface carrying the layer.

We calculate the field quantities due to such a layer by generalizing eqns. (2.36) and (2.37) in a manner similar to our generalizations of the field of a point charge. Thus

$$V = \iint \frac{\tilde{\omega}(\hat{v}\cdot r)\,\mathrm{d}S}{4\pi\varepsilon_0 r^3} = \iint \frac{\tilde{\omega}}{4\pi\varepsilon_0}\,\hat{v}\cdot\mathbf{grad}_P\left(-\frac{1}{r}\right)\mathrm{d}S$$

$$= \iint \frac{\tilde{\omega}}{4\pi\varepsilon_0}\,\hat{v}\cdot\mathbf{grad}_R\left(\frac{1}{r}\right)\mathrm{d}S.$$

(The suffixes P, R denote the point of differentiation, and the integrals are

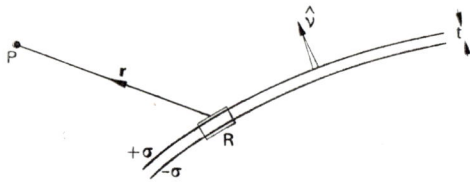

Fig. 2.28

taken over the surface carrying the double layer.) If $\tilde{\omega}$ is uniform, we can evaluate the last integral, thus

$$V = \frac{\tilde{\omega}}{4\pi\varepsilon_0} \iint \hat{v}\cdot\mathbf{grad}_R\left(\frac{1}{r}\right)\mathrm{d}S = \frac{\tilde{\omega}}{4\pi\varepsilon_0}\,\omega_P, \qquad (2.41)$$

where ω_P denotes the solid angle subtended at P by the perimeter of the surface carrying the double layer. The solid angle is taken positively when a radius drawn *from* P to the surface, inside the angle, meets the positive face of the layer first. We do not attempt here to use the integral form giving $\mathbf{D}$ or $\mathbf{E}$ which is

$$\mathbf{D} = \frac{1}{4\pi}\iint\left\{-\frac{\tilde{\omega}\hat{v}}{r^3} + \frac{3\tilde{\omega}(\hat{v}\cdot r)r}{r^5}\right\}\mathrm{d}S = \varepsilon_0\mathbf{E}.$$

It is easier to take the gradient of V as obtained from eqn. (2.41).

To investigate the discontinuities produced by the double layer we consider the discontinuities of each of the single layers in Fig. 2.28 and take the limit as before. We denote field quantities in the narrow space between the layers by a suffix zero. Hence, from eqn. (2.28) we obtain

$$\hat{v}\cdot(\mathbf{D}_+ - \mathbf{D}_0) = \sigma, \quad \hat{v}\cdot(\mathbf{D}_0 - \mathbf{D}_-) = -\sigma.$$

Therefore
$$\hat{v}\cdot(\mathbf{D}_+ - \mathbf{D}_-) = 0. \qquad (2.42)$$

Similarly for the tangential components of E,

$$\hat{v} \times (E_+ - E_-) = 0. \tag{2.43}$$

We use the same value for D_0 (or E_0) in each of the equations above leading to eqn. (2.42) since the field between the surfaces is large but effectively uniform.

In considering the discontinuity in the potential function we note that V is continuous across each single layer. However, the field arising from the single layers, when t is small, dominates all other fields, such as those from remote charges. Since the electric field in the space is $-(\sigma/\varepsilon_0)\hat{v}$, there is a change of potential between the layers equal to $\sigma t/\varepsilon_0$, and we deduce that

$$V_+ - V_- = \lim (\sigma t/\varepsilon_0) = \tilde{\omega}/\varepsilon_0. \tag{2.44}$$

Example 1. Alternatively, we can deduce eqn. (2.44), in the case of an open surface, by using the result (2.41) to follow the change in V as the field point moves from P_- to P_+, as shown in Fig. 2.29. At P_- the solid angle subtended by the perimeter is $-\omega$, ($\omega > 0$). The solid angle is negative according to the convention of eqn. (2.41). At P_+ the solid angle is

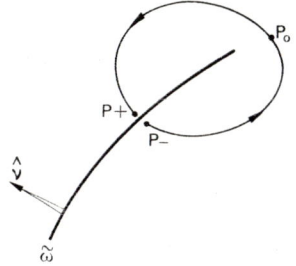

FIG. 2.29

$4\pi - \omega$. As the field point follows the path indicated the solid angle becomes zero at a point such as P_0, from there to P_+ it increases. Hence the potential increases steadily from V_-, to zero, to V_+, where

$$V_- = -\frac{\tilde{\omega}\omega}{4\pi\varepsilon_0}, \quad V_+ = \frac{\tilde{\omega}(4\pi - \omega)}{4\pi\varepsilon_0}.$$

Therefore
$$V_+ - V_- = \tilde{\omega}/\varepsilon_0.$$

Example 2. A corollary of eqn. (2.41) is that the field produced by a *closed uniform double layer* is zero. Since the surface has no perimeter, $\omega = 0$ for all external points, and $\omega = -4\pi$ for all internal points. Hence

$$V = 0 \quad \text{outside},$$
$$V = -\tilde{\omega}/\varepsilon_0 \quad \text{inside}.$$

Both these values are constant so that E and D are zero everywhere inside and outside the double layer.

2.12 Volume distributions of dipoles: polarization

A region of space contains *polarization* when every element of that region behaves as a dipole. If a small element of volume v behaves as a dipole of moment μ, then we define the (*intensity of*) *polarization* at a point within the element by the vector $\boldsymbol{P}$, where

$$\boldsymbol{P} = \lim_{v \to 0} (\boldsymbol{\mu}/v). \tag{2.45}$$

We can regard a polarized region as the limiting case of two regions of volume charge which have been superimposed in such a way that two volume elements v, carrying equal and opposite volume densities $\pm \varrho$, originally separated by a displacement $\boldsymbol{l}$ have coincided and $\varrho \to \infty$, $l \to 0$ in such a way that

$$\boldsymbol{P} = \lim \varrho \boldsymbol{l},$$

where $|\boldsymbol{P}|$ is finite.

We use this method to obtain some simple results concerning uniformly polarized bodies of simple shapes.

Example 1. We consider two identical cuboids, the one carrying uniform charge density ϱ with its centre at C_+, the other carrying a uniform charge density $-\varrho$ with its centre at C_-. [See Fig. 2.30 (i).] The displacement $C_- C_+$ is $\boldsymbol{l}$ in a direction parallel to an edge. Hence two corresponding volume elements, one from each block of charge, are also separated by $\boldsymbol{l}$. We now make $\varrho \to \infty$, $l \to 0$ so that $\boldsymbol{P} = \lim \varrho \boldsymbol{l}$. Then every pair of such volume elements v becomes a dipole of moment $\boldsymbol{P}v$, being the limit of charges $\pm \varrho v$. However, at every stage of this limiting process the equal and opposite charges $\pm \varrho$ in the main body of

Fig. 2.30

the volume cancel each other out, leaving only a layer of positive charge on one end face, and a layer of negative charge on the other end face. In a small area α of the right-hand end face the charge is $\varrho \alpha l$, a charge $-\varrho \alpha l$ being situated on a similar area of the other end face. In the limit these two charges do not vanish but become $\pm |\boldsymbol{P}| \alpha$. Hence the uniformly polarized rectangular block is equivalent to surface charges on the end faces. The surface density of either of these charges can be written as $\sigma' = \boldsymbol{P} \cdot \hat{\boldsymbol{\nu}}$ at any point (including the side faces) when $\hat{\boldsymbol{\nu}}$ is the outward normal to the boundary surface of the polarization.

Example 2. As a second example we consider two uniform spherical volume charges superimposed in a similar manner. The limit is a uniformly polarized sphere, whose electrostatic effects are equivalent to a certain surface layer of simple charge.

In the direction shown by θ in Fig. 2.30 (ii) the thickness of the layer is $l\cos\theta$; hence the charge adjacent to a small area α at this point is $\alpha\varrho l\cos\theta$. In the limit this becomes a surface charge of density $|\boldsymbol{P}|\cos\theta$. (This applies to both the positive and the negative faces on taking account of the sign of $\cos\theta$.) This surface density can again be written $\sigma' = \boldsymbol{P}\cdot\hat{\boldsymbol{v}}$ and is equivalent to the uniformly polarized sphere.

In the case of the sphere we can easily find the field produced by this distribution. For field points outside both distributions of simple charge the field is the same as that due to two point charges (see examples 4 and 5, pp. 43–45),

$$\tfrac{4}{3}\pi a^3 \varrho \quad \text{at} \quad C_+, \quad -\tfrac{4}{3}\pi a^3 \varrho \quad \text{at} \quad C_-.$$

In the limit these are equivalent to a dipole of moment

$$\boldsymbol{\mu} = \lim \tfrac{4}{3}\pi a^3(\varrho l) = \tfrac{4}{3}\pi a^3 \boldsymbol{P}.$$

For a field point P inside both spheres we use eqn. (1) of example 5, p. 44, which gives

$$\varepsilon_0 \boldsymbol{E} = -\tfrac{1}{3}\varrho\overrightarrow{C_- P} + \tfrac{1}{3}\varrho\overrightarrow{C_+ P} = -\tfrac{1}{3}\varrho\overrightarrow{C_- C_+} = -\tfrac{1}{3}\varrho l.$$

In the limit this becomes $\varepsilon_0 \boldsymbol{E} = -\tfrac{1}{3}\boldsymbol{P}$.

Example 3. The field in a cylindrical cavity, with its axis parallel to the direction of the polarization (Fig. 2.31).

The limiting process used in the above examples shows that the faces of the cylinder perpendicular to $\boldsymbol{P}$ carry charges which have surface densities $\pm|\boldsymbol{P}|$. Hence the field inside the cavity is that contributed from distant charges, including any other boundary surfaces of $\boldsymbol{P}$, together with the local contribution from the faces.

If we choose the cavity to have a radius negligible compared to its length—a *needle-shaped* cavity—the contribution from the local surfaces of the cavity is negligible and any field in the cavity must be due to distant charges only.

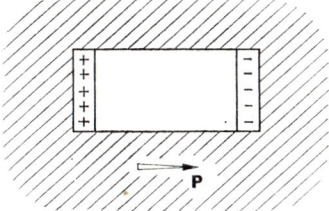

Fig. 2.31

If we choose the cavity to have a radius large compared with its axial length, for points inside the cavity the local contribution to $\boldsymbol{D}$ from the faces is effectively that from two equal and opposite surface densities on infinite surfaces. In this case

$$\boldsymbol{D} = \sigma\hat{\boldsymbol{v}} = \boldsymbol{P}.$$

When the polarization of a region is not uniform we treat each volume element $d\tau$ as a dipole of moment $\boldsymbol{P}d\tau$ and sum the contributions by integration. Hence [cf. eqn. (2.37)] the expressions for the field quantities V, $\boldsymbol{D}$, $\boldsymbol{E}$

corresponding to eqns. (2.14) and (2.15) are

$$D = \iiint \left\{ -\frac{P}{4\pi r^3} + \frac{3(P \cdot r)r}{4\pi r^5} \right\} d\tau = \varepsilon_0 E, \qquad (2.46)$$

$$V = \iiint \frac{P \cdot r}{4\pi \varepsilon_0 r^3} d\tau. \qquad (2.47)$$

In these integrals P is a function of $\{\xi\ \eta\ \zeta\}$, the coordinates of the volume element, and

$$r = i(x-\xi) + j(y-\eta) + k(z-\zeta), \quad r = |r|,$$

where $\{x, y, z\}$ are the coordinates of the field point.

For field points external to the polarized region r never vanishes and the integrals are continuous functions of the position of the field point, and may be differentiated. Hence, as usual, $E = -\operatorname{grad} V$.

For field points inside the polarized region, V is an integral of the type $I(2)$, and so is convergent, but D and E are of the type $I(3)$. The latter are indeterminate, and so we must modify our definitions corresponding to eqn. (2.46) for internal points; however, first we consider an alternative form for V.

For both internal and external points we have

$$V = \frac{1}{4\pi\varepsilon_0} \iiint P \cdot \operatorname{grad}_P \left(-\frac{1}{r}\right) d\tau.$$

Since P is a function only of the coordinates $\{\xi\ \eta\ \zeta\}$ of the source point Q at the element $d\tau$, it is independent of the coordinates $\{x\ y\ z\}$ of the field point P. Therefore

$$P \cdot \operatorname{grad}_P \left(-\frac{1}{r}\right) = P \cdot \operatorname{grad}_Q \left(\frac{1}{r}\right) = \operatorname{div}_Q \left(\frac{P}{r}\right) - \frac{1}{r} \operatorname{div}_Q P.$$

Therefore $\quad V = \dfrac{1}{4\pi\varepsilon_0} \iiint \left\{ \operatorname{div}_Q \left(\dfrac{P}{r}\right) - \dfrac{1}{r} \operatorname{div}_Q P \right\} d\tau,$

i.e. $\quad V = \dfrac{1}{4\pi\varepsilon_0} \iiint \dfrac{(-\operatorname{div} P)}{r} d\tau + \dfrac{1}{4\pi\varepsilon_0} \iint \dfrac{P \cdot \hat{v}}{r} dS, \qquad (2.48)$

where we have used the divergence theorem to obtain the surface integral which is taken over the surface bounding the polarized region. This gives V as the potential arising from a density of simple charge $\varrho' = -\operatorname{div} P$ and a surface charge on the boundary $\sigma' = P \cdot \hat{v}$. This distribution of charge is called *Poisson's equivalent distribution*. We obtained special cases of this in the examples above where, since **P** was uniform, $\varrho' = 0$. Poisson's equivalent charge produces the same field as the polarization at external field points, and we now use this to resolve the indeterminacy for internal points. The transformation from eqn. (2.47) to eqn. (2.48) is called *Poisson's transformation*.

When we investigate the field at internal points we consider the region to have a small cavity which excludes the field point *P* from the region of integration (Fig. 2.32). With this modified region of integration the integral of eqn. (2.46) has a definite value, but this value tends to a limit which depends on the shape of the cavity when the dimensions of the cavity are reduced to zero. These different limiting values arise because of the contribution from the

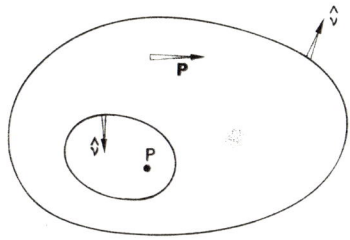

Fig. 2.32

Poisson charge $\mathbf{P} \cdot \hat{\mathbf{v}}$ on the walls of the cavity. Because the integrals (2.47) and (2.48) for *V* are both convergent, the value of *V* is not affected by the shape of the cavity and we may use either expression.

We resolve the indeterminacy for **D** and **E** by defining them in terms of the values obtained by using special-shaped cavities, and *we use a different-shaped cavity for **E** from that used to define **D***. The cavities we use are small (tending to zero in the limit) so that we may take **P** to be uniform in the volume immediately surrounding the cavities and apply the results obtained in example 3, p. 65. When the cavity has a "needle" shape, i.e. its radius is small compared with its length, the local contribution is negligible. When the cavity has a disc (or coin) shape, i.e. its axial length is small compared with its radius, the cavity walls make the maximum contribution.

We *define* the vector **E** at an internal point by the limiting value of **E** in a needle-shaped cavity. In this case **E** arises from the equivalent Poisson charge σ' on all boundary surfaces and the equivalent volume charge ϱ', with no local contribution. Hence, using the Poisson charge in eqn. (2.14),

$$E = \frac{1}{4\pi\varepsilon_0} \iiint \frac{(-\operatorname{div} \mathbf{P})\mathbf{r}}{r^3} \, d\tau + \frac{1}{4\pi\varepsilon_0} \iint \frac{(\mathbf{P} \cdot \hat{\mathbf{v}})\mathbf{r}}{r^3} \, dS, \qquad (2.49)$$

the surface integral being taken over all boundary surfaces of the polarized region.

We *define* the vector D to be the limiting value of D in a disc-shaped cavity when the local contribution is a maximum. The result of example 3, p. 65, then shows that

$$D = \frac{1}{4\pi} \int\int\int \frac{(-\operatorname{div} P)r}{r^3} \, d\tau + \frac{1}{4\pi} \int\int \frac{(P \cdot \hat{\nu})r}{r^3} \, dS + P. \qquad (2.50)$$

Since our original definition of E was in terms of the force on a point charge it is "sensible" to use a long narrow cavity for defining E. Since our original definition of D was in term of the displacement across an area (or the charges on the faces of two discs), it is "sensible" to use a disc-shaped cavity to define D. Moreover, with the addition of being measured in an appropriate cavity, the new definitions are based on the same measurements as the original definitions. From eqns. (2.49) and (2.50) we obtain the important relation

$$D = \varepsilon_0 E + P. \qquad (2.51)$$

At those points where $P = 0$ this equation reduces to eqn. (2.10) and so may be regarded as the generalization of eqn. (2.10). We also see that, by taking the gradient of V, given by eqn. (2.48), we obtain the usual relation $E = -\operatorname{grad} V$ for both internal and external points. A comparison of eqn. (2.15) with eqn. (2.48) shows that we can write down an equation corresponding to eqn. (2.25), viz.

$$\varepsilon_0 \nabla^2 V = -\varrho' = \operatorname{div} P.$$

Therefore $\operatorname{div} D = -\operatorname{div}(\varepsilon_0 \operatorname{grad} V) + \operatorname{div} P = -\varepsilon_0 \nabla^2 V + \operatorname{div} P = 0.$ (2.52)

If there is a volume distribution of simple charge ϱ present, in addition to the polarization, then, by the principle of superposition, eqn. (2.52) becomes $\operatorname{div} D = \varrho$, which is eqn. (2.24). Hence we may regard eqn. (2.24) as true in all circumstances.

Now we consider what happens at a discontinuity of P where there is no surface charge σ. The equivalent Poisson surface distribution is $(\hat{\nu} \cdot P_-)$ from the polarization inside the surface of discontinuity and $(-\hat{\nu} \cdot P_+)$ from that outside. (See Fig. 2.33.) Hence there is an effective surface density

$$\sigma' = -\hat{\nu} \cdot (P_+ - P_-)$$

together with the volume densities

$$\varrho'_- = -\operatorname{div} P_- \text{ inside}, \quad \varrho'_+ = -\operatorname{div} P_+ \text{ outside}.$$

Equation (2.49) shows that $\varepsilon_0 E$ is derived from ϱ', σ' in exactly the same way as D (or $\varepsilon_0 E$) is derived from ϱ and σ in eqn. (2.14). Hence we may quote

§ 2.12 THE INVERSE SQUARE LAW IN ELECTROSTATICS

eqns. (2.28) and (2.29) and obtain

$$\hat{v} \cdot (\varepsilon_0 E_+ - \varepsilon_0 E_-) = -\hat{v} \cdot (P_+ - P_-).$$

Therefore
$$\hat{v} \cdot \{(\varepsilon_0 E_+ + P_+) - (\varepsilon_0 E_- + P_-)\} = 0,$$

i.e.
$$\hat{v} \cdot (D_+ - D_-) = 0. \tag{2.53}$$

Hence, eqn. (2.28) still applies even in the presence of polarization. Similarly, eqn. (2.29),

$$\hat{v} \times (E_+ - E_-) = 0,$$

still applies. [This is another reason for using the two equations (2.28) and (2.29) instead of the single relation (2.27); the latter does not hold when polarization is present.]

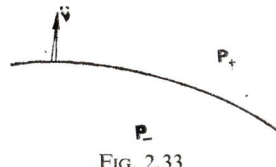

Fig. 2.33

We are now in a position to sum up all the results of this chapter in a few important formulae. The sources of the field are simple charges ϱ, σ, Q, together with dipoles P and $\tilde{\omega}$. The equations are:

$$\left. \begin{array}{ll} \text{Conservation of energy:} & \text{curl } E = 0, \quad E = -\text{grad } V; \\ \text{Inverse square law:} & \text{div } D = \varrho, \quad \varepsilon_0 \nabla^2 V = -\varrho; \\ \text{Relation of field vectors:} & D = \varepsilon_0 E + P. \end{array} \right\} \tag{2.54}$$

The corresponding surface relations are

$$\begin{array}{ll} \hat{v} \times (E_+ - E_-) = 0, & V_+ - V_- = \tilde{\omega}/\varepsilon_0, \\ \hat{v} \cdot (D_+ - D_-) = \sigma, & (\partial V/\partial n)_+ - (\partial V/\partial n)_- = -\sigma/\varepsilon_0. \end{array} \tag{2.55}$$

Exercises 2.12

1. Determine the potential at any point *in vacuo* of a line doublet of moment M per unit length. Discuss the nature of the equipotential surfaces, and find the equations of the lines of force.
2. Show that if we do not assume V is continuous in § 2.9(4), p. 56, then the integral $\iint_{S_k} V \frac{\partial}{\partial n}\left(\frac{1}{r}\right) dS$ becomes $\iint_K (V_- - V_+) \frac{\partial}{\partial n}\left(\frac{1}{r}\right) dS$, and so leads to the discontinuity relation (2.44).

3. Find that part of the field inside a spherical cavity in a uniformly polarized region arising from the contribution made by the walls of the cavity.
4. A region bounded by a surface S is occupied by a distribution of electric dipoles of volume density P. Show that the field at any point outside S is the same as the field due to certain distributions of charge over the surface S and throughout the volume bounded by S.
A small cube with edges parallel to the unit vectors i, j, k is removed from the dielectric. Prove that the electric intensity at the centre of the cube is

$$\tfrac{1}{3}\varepsilon_0\{(P\cdot i)i+(P\cdot j)j+(P\cdot k)k\},$$

plus the field due to the charges on S and in the remainder of the solid.

Miscellaneous Exercises II

1. Give careful sketches indicating the most significant features of the lines of force and equipotential lines in the following cases:
    (i) Two equal point charges of the same sign.
    (ii) Two equal and opposite point charges.
    (iii) Two charges $4q$ and $-q$.
    (iv) A point of charge placed in a uniform field of force.
2. Point charges of $4q$, $-q$, $4q$ are placed at the points $(-a, 0, 0)$, $(0, 0, 0)$, $(a, 0, 0)$ respectively. Prove that
    (i) there is a ring of points of equilibrium along the circle $x = 0$, $y^2+z^2 = \tfrac{1}{3}a^2$;
    (ii) the lines of force through these points leave the positive charges at an angle $\cos^{-1}(\tfrac{3}{4})$;
    (iii) the asymptotes to the lines of force which go to infinity pass through the origin.
Sketch roughly the system of lines of force in a plane through the line of charges.
3. A particle of charge e and mass m is released from rest at a point on the axis of an electrical dipole of strength μ, at distance a from it. Write down the equations for conservation of energy and for the radial acceleration of the particle. By eliminating the angular motion, show that after time t the particle is at distance r from the dipole, where

$$r^2 = a^2 + \frac{\mu e t^2}{2\pi m \varepsilon_0 a^2}.$$

If the particle, instead of being released from rest, is projected from the same point with velocity V in a direction at right angles to the axis of the dipole, and if μ has the particular value

$$\mu = -\frac{2\pi m \varepsilon_0 a^2 V^2}{e},$$

show that the particle moves in a circular arc of radius a in the same way as a simple pendulum of length $2ga^2/V^2$ swinging through an angle π.

CHAPTER 3

CONDUCTORS IN THE ELECTROSTATIC FIELD

3.1 The property of a conductor

The formulae (2.14) and (2.15) of Chapter 2 give complete information concerning an electrostatic field arising from a *known* distribution of simple charge. In principle, when conductors are present, charges can move on the conductors, but not leave the conductors, as the charges take up their equilibrium positions. These equilibrium positions must therefore depend upon the field prevailing; nevertheless the field given by eqn. (2.14) depends upon the positions of the charges. It is this interdependence of charge distribution and field, arising because charges can move on conductors, which constitutes the basic problem of electrostatics. This problem we consider now.

We draw certain conclusions relating the field and the charges on the conductors from the fundamental property that charges can move on a conductor.

(1) *There can be no field within the material of a conductor.* Since a charge can move in the conductor when acted upon by a force, the charge cannot be in equilibrium unless $E = 0$.

(2) From this result we deduce that *any charge on a conductor must be situated on the surface* as a thin layer, in fact, as a surface charge. We establish this result by drawing an arbitrary Gauss surface G lying entirely *within* the material of the conductor; because E (and therefore D) is zero at all points of G, this surface encloses no charge. Since G is arbitrary we conclude that there can be no charge at any point *inside* the conducting material. The result follows.

(3) Because there is no field inside a conductor, no work is required to move a test charge anywhere in the conductor, and so *all points of the conductor are at the same potential.* Hence, in an electrostatic field the *surface of a conduc-*

tor is an equipotential surface which carries a surface density of charge σ; the latter, in general, is not uniform.

(4) Because the boundary surface of a conductor is an equipotential surface of the electrostatic field and the field lines are orthogonal trajectories of these equipotential surfaces, the field lines at a point in the space immediately outside the conductor must lie along the normals to the conductor near that point. Also, the discontinuity in $\hat{\nu}\cdot\boldsymbol{D}$ corresponding to the surface charge σ implies that at any point immediately outside a conductor the field is given by

$$\boldsymbol{D} = \hat{\nu}\sigma = \varepsilon_0\boldsymbol{E}, \qquad (3.1)$$

or

$$\frac{\partial V}{\partial \nu} = -\frac{\sigma}{\varepsilon_0}. \qquad (3.2)$$

Here the differentiation is taken along the normal drawn into the field. Also

$$V = \text{constant} \qquad (3.3)$$

on the surface, and inside the material, of a conductor.

3.2 Some general theorems

When we discuss the general theorems of electrostatics it is essential that we consider the most general possible arrangements of charges and conductors. Then any particular fields examined later can be taken as special cases of this general arrangement.

As pointed out earlier, electric charge is not rigidly fixed unless it is situated on an insulator. But this insulator may itself modify the field (see Chapter 4). Consequently, in the absence of solid dielectrics, a rigid distribution of charge, especially of volume charge, is unlikely to be realized in practice, even approximately. The most usual arrangement is a field in which a number of conductors carry charges, or are maintained at stated potentials and are situated in a vacuum or, more likely, in air. (The difference between air and a vacuum for this purpose is negligible in most cases.) Nevertheless, we include the possibility of rigid distributions of charge in establishing the general theorems, although the usual applications are to fields established in the space between a number of charged conductors.

The earth is often considered to be one of the conductors of a field, even when not mentioned explicitly, and is assumed to be always at zero potential no matter what charge resides on it. Any other conductor put into metallic connection with the earth immediately takes up a potential zero by drawing

§ 3.2   CONDUCTORS IN THE ELECTROSTATIC FIELD   73

charge, if necessary, from the earth onto itself along the conducting connection. If this connection is subsequently broken, this charge remains on the conductor. (This process is the basis of a very common method of charging a conductor in elementary practice, e.g. the electrophorus.) Sometimes the earth is regarded as a large conducting plane at a finite distance from other conductors. More often the earth is regarded as that conductor at infinity on which can end all those unit tubes, arising from (or ending on) conductors in the field, which cannot end on (or start from) another conductor at a finite distance. Thus, the unit tubes which arise from an "isolated" point charge radiate uniformly from this charge and end on "the earth" at infinity. This use of "the earth" as a conductor at infinity can be compared with the use of the "point at infinity" in geometry. The "earth" has a conventional diagrammatic notation; see, for example, Fig. 3.1, p. 76.

To overcome the difficulty pointed out in § 3.1, we reduce the general problem of electrostatics to finding the function V for all points of the field. To find V we refer to eqn. (2.25) and formulate the problem as follows.

The field is in a region, which may or may not go to infinity, having a finite number of conductors S_i as its finite boundaries. The problem is to find a function V defined in this region and subject to the following conditions:

(1) V satisfies Poisson's equation, eqn. (2.25),

$$\nabla^2 V = -\varrho/\varepsilon_0,$$

where ϱ is the (rigid) volume distribution of charge. If $\varrho = 0$, V satisfies Laplace's equation $\nabla^2 V = 0$.

(2) V satisfies the discontinuity relations, eqn. (2.33),

$$\left(\frac{\partial V}{\partial \nu}\right)_+ - \left(\frac{\partial V}{\partial \nu}\right)_- = -\frac{\sigma}{\varepsilon_0}$$

at all points of a surface carrying a rigid surface distribution of charge σ. In the absence of double layers (which seldom arise in practice) the potential function is continuous, eqn. (2.32), across layers of surface charge.

(3) V has a singularity at every point charge Q_l such that

$$V = \frac{Q_l}{4\pi\varepsilon_0 s} + O(1), \quad |\mathrm{grad}\, V| = \frac{Q_l}{4\pi\varepsilon_0 s^2} + O(1), \quad s \to 0,$$

where s is the distance of the field point from the singularity Q_l.

(4) On each (finite) conducting boundary S_i of the field,

$$V = V_i = \text{constant}, \quad \text{and} \quad \oint_{S_i} \frac{\partial V}{\partial \nu}\, dS = -\frac{Q_i}{\varepsilon_0}.$$

These are equivalent to eqns. (3.2) and (3.3).

(5) If the region of the field goes to infinity,
$$V = O\left(\frac{1}{R}\right), \quad |\operatorname{grad} V| = O\left(\frac{1}{R^2}\right) \quad \text{as} \quad R \to \infty,$$
i.e. V and its derivatives obey the standard boundary conditions at infinity.

The problem is therefore one of solving a second-order partial differential equation for V subject to certain boundary and continuity conditions. It is condition (4) which takes account of the mobility of charge on a conducting boundary.

When we specify the boundary conditions under heading (4), for each conductor there are two possibilities. Either we know the potential V_i of that conductor, and can subsequently calculate the charge Q_i when V has been determined, or we know the total charge Q_i [i.e. the integral $\oiint (\partial V/\partial \nu)\, dS$ but not the detailed values of $\partial V/\partial \nu$] but do not know the potential; in this latter case we can write $V = V_i$ (a constant) at all points of the conductor but we do not know the value V_i.

When the potential function V has been determined we use eqn. (3.2) to find the actual distribution of the charge over the conducting boundaries. In the important case of a field arising from a number of charged conductors in a vacuum (or air) the differential equation is Laplace's equation, i.e. V is a *harmonic function* and we refer only to the boundary conditions (4) and (5).

The first important general theorem we consider is the Uniqueness theorem, which states that there is only one solution to the problem summarized by (1)–(5) above. This is to be expected on physical grounds—no single arrangement of charges would give rise to two possible fields. In *C.M.*, vol. 4, § 1.8, uniqueness theorems were discussed in abstract. In that discussion the boundary conditions were almost identical with those above except for eqn. (4). In the abstract discussion we assumed that either V or its normal derivative $\partial V/\partial \nu$ was specified at every point of the boundary. In the present case this is replaced by the condition (4) above. The solution is still unique.

We assume two possible solutions V, V' and write $U = V - V'$. Then U satisfies the following conditions.

(1) Since $\nabla^2 V = -\varrho/\varepsilon_0 = \nabla^2 V'$,
$$\nabla^2 U = 0,$$
i.e. U is harmonic *everywhere* in the field.

(2) Since at points of a surface carrying a rigid surface distribution of charge
$$\left(\frac{\partial V}{\partial \nu}\right)_+ - \left(\frac{\partial V}{\partial \nu}\right)_- = -\frac{\sigma}{\varepsilon_0} = \left(\frac{\partial V'}{\partial \nu}\right)_+ - \left(\frac{\partial V'}{\partial \nu}\right)_-,$$

§ 3.2 CONDUCTORS IN THE ELECTROSTATIC FIELD

it follows that
$$\left(\frac{\partial U}{\partial \nu}\right)_+ - \left(\frac{\partial U}{\partial \nu}\right)_- = 0$$

on all surfaces of discontinuity, i.e. $\partial U/\partial \nu$ is continuous.

(3) Since V and V' have the same singularities at the point charges, then U has no singularities at these points.

(4) On those boundary surfaces where the potential is given
$$U = V_i - V_i = 0,$$

and on those surfaces where the charge is given but the potential is unknown, $U = \text{constant}$ and
$$\oiint_{S_i} \frac{\partial U}{\partial \nu} \, dS = \oiint_{S_i} \left(\frac{\partial V}{\partial \nu} - \frac{\partial V'}{\partial \nu}\right) dS = \frac{Q_i}{\varepsilon_0} - \frac{Q_i}{\varepsilon_0} = 0.$$

(5) Where the region goes to infinity
$$U = O\left(\frac{1}{R}\right), \quad \left|\frac{\partial U}{\partial n}\right| = O\left(\frac{1}{R^2}\right).$$

The application of Green's theorem [*C.M.*, vol. 4, eqn. (1.57)] to U gives
$$\iiint \{U \nabla^2 U + (\nabla U)^2\} \, d\tau = \oiint_{S_0} U \frac{\partial U}{\partial n} \, dS - \sum_i \oiint_{S_i} U \frac{\partial U}{\partial n} \, dS,$$

where S_0 is a surface large enough to enclose all charges and conductors.

This equation reduces on applying the conditions above to
$$\iiint (\nabla U)^2 \, d\tau = 0.$$

Hence $\nabla U = 0$ everywhere and so $U = \text{constant}$. The value of the constant is zero since $U = 0$ on one of the boundaries (and/or at infinity). Therefore
$$V = V' \tag{3.4}$$

everywhere in the field.

(In the special case of a field inside a closed conducting boundary, and for which the potential is nowhere specified on any boundary, since the field does not go to infinity, we could have $U = \text{constant}$.)

This result is of great importance, for most electrostatic problems are solved, not by writing down a general solution, but by using symmetry and familiar solutions of simpler problems to suggest (by inspired guessing) a potential function which satisfies all the conditions. The uniqueness theorem then implies that this is the only solution.

Example 1. The field inside a closed hollow conductor which encloses no charge of either kind is zero.

This is a special case of the above general problem. In this case the function V has to satisfy

$$\nabla^2 V = 0 \text{ everywhere inside } S, \qquad (1)$$

where S is the inner surface of the conductor.

There are no discontinuities. (2)

There are no singularities. (3)

On S, $C = V$. (4)

The region does not include infinity since S is closed.

The function $V(x, y, z) = C$ satisfies all the conditions, and is therefore the solution. Hence we conclude that there is no field inside this hollow conductor and no surface density of charge on the inner surface.

The above result does not apply if the surface encloses equal and opposite amounts of charge. In this case at least one of eqns. (1), (2) or (3) cannot be satisfied, e.g. there may be two point charges $\pm Q$, i.e. two singularities, or there may be two regions, one with positive ϱ and another with negative ϱ, and so eqn. (1) is not satisfied.

This same result can be obtained by an argument using the properties of tubes of force. Suppose that there is a field inside the conductor. Since there is no charge of any kind to give a start or end to such tubes, these tubes must start and end on the inner surface of the conductor. But the two ends of a tube must be at different potentials and so two points of one and the same conductor must be at different potentials. This is impossible, and so our assumption of the presence of tubes of force must be rejected. *There can, therefore, be no field.*

Example 2. A spherical condenser consists of two concentric spheres insulated from each other. We consider the case in which the inner sphere, of radius a, carries a total charge Q and the outer sphere of radius b ($b > a$) is earthed (Fig. 3.1). This figure uses the conventional sign for connection to earth.

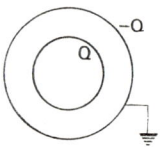

FIG. 3.1

A spherically symmetrical distribution of charge and a radial field between the conductors satisfies the conditions of the problem.

The general spherically symmetrical solution of Laplace's equation is

$$V = \frac{A}{r} + B.$$

When $r = b$, $V = 0$; therefore $0 = \frac{A}{b} + B$.

When $r = a$, $V = V_1$, say; therefore $V_1 = \frac{A}{a} + B$.

Therefore
$$V - V_1 = A\left(\frac{1}{r} - \frac{1}{a}\right).$$

On $r = a$, $\dfrac{\partial V}{\partial v} = \left(\dfrac{\partial V}{\partial r}\right)_{r=a} = -\dfrac{A}{a^2}$; therefore $\sigma_a = \dfrac{\varepsilon_0 A}{a^2}$.

On $r = b$, $\dfrac{\partial V}{\partial v} = -\left(\dfrac{\partial V}{\partial r}\right)_{r=b} = \dfrac{A}{b^2}$; therefore $\sigma_b = -\dfrac{\varepsilon_0 A}{b^2}$.

Therefore $Q_a = 4\pi\varepsilon_0 A = Q$, $Q_b = -4\pi\varepsilon_0 A = -Q$.

so that
$$V = V_1 + \frac{Q}{4\pi\varepsilon_0}\left(\frac{1}{r} - \frac{1}{a}\right), \quad V_1 = \frac{Q}{4\pi\varepsilon_0}\left(\frac{1}{a} - \frac{1}{b}\right).$$

Outside $r = b$, the solution is $V = 0$.
(This result can also be obtained by the use of Gauss's theorem and the spherical symmetry of the system.)

It can be seen from this result that the charge Q has attracted an equal and opposite charge $-Q$ on to the earthed conductor. This latter charge is frequently called an "induced charge".

Example 3. An insulated thin spherical shell of radius a, charge Q, potential V_1, is surrounded by a concentric thin insulated spherical shell of radius $2a$, charge $-2Q$, potential V_2, which in turn is surrounded by a concentric thin insulated spherical shell of radius $3a$, charge $3Q$, potential V_3. Find V_1, V_2, V_3, in terms of Q/a, and prove that
$$3V_3 = V_1 + 2V_2.$$

Figure 3.2 shows the radii and the charges on the respective surfaces of the spheres. The whole is spherically symmetrical.

The arrangement of the charges is arrived at by requiring the total charge on the inner and outer surface of each sphere to add up to that specified. There is no field in the region for which $r < a$, see example 1 above. In the region for which $a \leqslant r \leqslant 2a$ all the tubes arising from the positive charge Q residing on the sphere $r = a$ must end on the inner surface of the sphere $r = 2a$. Similarly in the region for which $2a \leqslant r \leqslant 3a$ all the tubes ending on

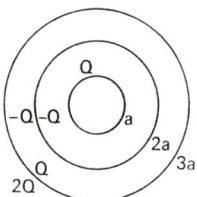

FIG. 3.2

the charge $-Q$ must start from a charge $+Q$ situated on the inner surface of the sphere $r = 3a$. The charges $-Q$ on the inner face of the sphere $r = 2a$ and $+Q$ on the inner face of the sphere $r = 3a$ are induced charges.

We can now write down the function V, in four parts, for the four regions thus:

	$0 \leqslant r \leqslant a$	$a \leqslant r \leqslant 2a$	$2a \leqslant r \leqslant 3a$	$3a \leqslant r$
$V(r) =$	V_1	$\dfrac{A_1}{r} + B_1$	$\dfrac{A_2}{r} + B_2$	$\dfrac{A_3}{r}$
$\dfrac{dV}{dr} =$	0	$-\dfrac{A_1}{r^2}$	$-\dfrac{A_2}{r^2}$	$-\dfrac{A_3}{r^2}$

We determine the constants A_1, A_2 from the boundary conditions relating surface charge to $\partial V/\partial v$, and B_1, B_2 from the continuity of V. We have put $B_3 = 0$ for $r \geqslant 3a$ since $V(r) \to 0$ as $r \to \infty$.

The surface densities of charge on the inner and outer surfaces of the spheres are

$$r = a \qquad r = 2a \qquad r = 3a$$

$$0, \; \frac{Q}{4\pi a^2}; \qquad -\frac{Q}{16\pi a^2}, \; -\frac{Q}{16\pi a^2}; \qquad \frac{Q}{36\pi a^2}, \; \frac{Q}{18\pi a^2}.$$

The direction of differentiation $\partial/\partial v$ in eqn. (3.2) is radial on each outer surface, and on each inner surface $\partial/\partial v = -d/dr$. Hence we obtain:

$$r = a: \qquad 0 = 0 \qquad\qquad -\frac{A_1}{a^2} = -\frac{Q}{4\pi\varepsilon_0 a^2};$$

$$r = 2a: \qquad +\frac{A_1}{4a^2} = \frac{Q}{16\pi\varepsilon_0 a^2} \qquad -\frac{A_2}{4a^2} = \frac{Q}{16\pi\varepsilon_0 a^2};$$

$$r = 3a: \qquad \frac{A_2}{9a^2} = -\frac{Q}{36\pi\varepsilon_0 a^2} \qquad -\frac{A_3}{9a^2} = -\frac{Q}{18\pi\varepsilon_0 a^2}.$$

Therefore $\qquad A_1 = \dfrac{Q}{4\pi\varepsilon_0}, \quad A_2 = -\dfrac{Q}{4\pi\varepsilon_0}, \quad A_3 = \dfrac{Q}{2\pi\varepsilon_0}.$

(We could also obtain these results by noting that in each region the field is effectively that of a point charge situated at the origin.

Q for $a \leqslant r \leqslant 2a$, $\;-Q$ for $2a \leqslant r \leqslant 3a$, $\;2Q$ for $3a \leqslant r$.)

Hence, we have, in the four regions,

$$V(r) = V_1, \quad \frac{Q}{4\pi\varepsilon_0 r} + B_1, \quad -\frac{Q}{4\pi\varepsilon_0 r} + B_2, \quad \frac{Q}{2\pi\varepsilon_0 r}.$$

We now determine the values of B_1, B_2 by noting that V has the values V_1, V_2, V_3 on the respective conductors.

Therefore $\qquad V_1 = \dfrac{Q}{4\pi\varepsilon_0 a} + B_1,$

$$V_2 = \frac{Q}{4\pi\varepsilon_0 \cdot 2a} + B_1,$$

$$V_3 = -\frac{Q}{4\pi\varepsilon_0 \cdot 3a} + B_2 = \frac{Q}{2\pi\varepsilon_0 \cdot 3a}.$$

Therefore $\qquad B_1 = 0, \quad B_2 = \dfrac{Q}{4\pi\varepsilon_0 a},$

$$V_1 = \frac{Q}{4\pi\varepsilon_0 a}, \quad V_2 = \frac{Q}{8\pi\varepsilon_0 a}, \quad V_3 = \frac{Q}{6\pi\varepsilon_0 a},$$

and we easily verify that
$$3V_3 = V_1 + 2V_2.$$

We give further examples in later sections using the uniqueness theorem and other techniques for solution of electrostatic problems.

3.3 Systems of conductors: capacitance

In this section we consider the simple case of the electrostatic field arising from a number of charged conductors in a vacuum (or air). We assume that there is no rigid distribution of volume or surface charge, and, in general, that no point charges are present. When point charges are included, we regard them as limiting cases of charged conductors with vanishingly small linear dimensions. Furthermore, we are not concerned with the field quantities so much as with the total charge on, and the potential of, a conductor.

The results obtained in this section depend essentially on the principle of superposition which is expressed by the fact that Laplace's equation and the boundary conditions on V are linear in the potential V.

When a system consists of one isolated conductor only (and the earth at infinity) this linear relation between the charge Q and the potential V of the conductor leads to

$$Q = CV. \qquad (3.5)$$

The coefficient C is independent of Q and V and can depend only upon the shape of the conductor. It measures the *capacitance* of the conductor; the larger the capacitance of a conductor the larger the charge it carries for a given potential. The capacitance is measured in farads (F) and is the charge which must be placed on the isolated conductor in order to raise its potential by one volt (i.e. 1 farad = 1 coulomb/volt). The farad is a large unit and in practice capacitance is measured in microfarads (1 $\mu F = 10^{-6}$ F).

When the system consists of a number of conductors S_i, $i = 1, 2, \ldots, N$, the charge Q_i on any one is a linear function of the potentials, $V_1, V_2, \ldots, V_N$, thus

$$\left. \begin{array}{l} Q_1 = C_{11}V_1 + C_{12}V_2 + \ldots + C_{1N}V_N, \\ Q_2 = C_{21}V_1 + C_{22}V_2 + \ldots + C_{2N}V_N, \\ \ldots\ldots\ldots\ldots\ldots\ldots\ldots\ldots\ldots\ldots\ldots \\ Q_N = C_{N1}V_1 + \ldots \quad\quad + C_{NN}V_N, \end{array} \right\} \quad \text{i.e. } Q_i = \sum_{j=1}^{N} C_{ij}V_j, \qquad (3.6)$$

$$i = 1, 2, 3, \ldots, N,$$

a relation which is very conveniently expressed in matrix form

$$\mathbf{Q} = \mathbf{CV}, \quad \mathbf{Q} = \{Q_i\}, \quad \mathbf{C} = (C_{ij}), \quad \mathbf{V} = \{V_i\}. \qquad (3.7)$$

The scalars in eqn. (3.5) are replaced either by column vectors or by a square capacitance matrix.

The elements on the principal diagonal are called *coefficients of capacitance* and those off the diagonal are called *coefficients of influence* (or of *induction*).

The values of the coefficients C_{ij} must depend upon the shapes and relative positions of the conductors but not upon their charges or potentials. We obtain an interpretation of the element C_{ij} by considering the special case in which conductor K is raised to unit potential and the others are maintained at zero (earth) potential. The conductors S_i, except S_K, must be kept at zero potential by a metallic connection to earth. Hence no tube of force can originate on S_i and end on the earth or on another conductor S_j. Since S_K is at a higher potential than all the other conductors, no tube of force in the field can end on S_K. We therefore have the situation illustrated in Fig. 3.3; the tubes originating on S_K must end either on one of the other conductors S_i or

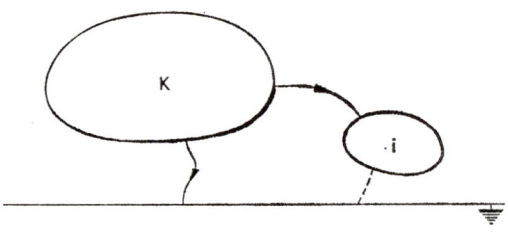

FIG. 3.3

on earth. We conclude from this that the charges on every conductor such as S_i are all negative and are together numerically less than (or perhaps equal to) the charge on S_K. But these charges are given by the coefficients in column K of the matrix C of eqn. (3.7). The charges on the conductors are C_{1K}, $C_{2K}, \ldots, C_{iK}, \ldots, C_{NK}$, and the above discussion shows that

$$C_{KK} + \sum_{i \neq K} C_{iK} \geq 0, \quad C_{iK} \leq 0 \quad \text{for} \quad i \neq K. \tag{3.8}$$

The off-diagonal elements C_{ij} therefore measure the effect of the electrostatic influence (or induction) of one conductor on another.

Electrostatic screening

If the conductors are so arranged that all lines of force originating (or ending) on S_J must end (or orginate) on S_K then S_K is said to *screen* S_J. This is called *electrostatic screening* and is illustrated in Faraday's ice-pail experiment. Screening is important practically when it is necessary to prevent some electrical apparatus from being disturbed by external electric influences and is achieved by mounting the apparatus inside a chamber with conducting walls, usually a wire mesh; the walls are also usually connected to earth

§ 3.3 CONDUCTORS IN THE ELECTROSTATIC FIELD

(though this is not theoretically necessary). Screening of this sort is also important in the design of radio circuits in which certain components must be protected from interference by other parts of the circuit, or from external sources.

Electrostatic screening of S_J by S_K is shown in eqn. (3.6) when $C_{iJ} = 0$ for $i \neq J$ or K, i.e. the coefficient of influence between S_J and any conductor other than S_K is zero. Let us consider the situation when S_J, S_K are at zero potential and another conductor S_i is raised to unit potential. All the tubes in the field originate on S_i, but all the tubes which can end on S_J must originate on S_K because of the screening, so that in this situation no tubes can end on S_J, which therefore carries no charge, i.e.

$$C_{iJ} = 0 \quad (i \neq J \text{ or } K).$$

Screening a conductor is usually done by enclosure in a hollow conductor; a capacitor or condenser (see p. 84) is an arrangement in which screening also takes place.

Because of the linear relation between charge and potential we could, in general, instead of eqn. (3.6), write down

$$\left.\begin{aligned} V_1 &= p_{11}Q_1 + p_{12}Q_2 + \ldots + p_{1N}Q_N, \\ V_2 &= p_{21}Q_1 + p_{22}Q_2 + \ldots + p_{2N}Q_N, \\ &\cdots\cdots\cdots\cdots\cdots\cdots\cdots\cdots\cdots\cdots\cdots \\ V_N &= p_{N1}Q_1 + \ldots \quad + p_{NN}Q_N, \end{aligned}\right\} \quad \text{i.e. } V_i = \sum_{j=1}^{N} p_{ij}Q_j, \quad (3.9)$$

$$i = 1, 2, 3, \ldots, N.$$

This is equivalent to solving eqns. (3.6) for V_i in terms of Q_i, or inverting the matrix eqn. (3.7) to give

$$V = C^{-1}Q, \quad Q = \{Q_i\}, \quad V = \{V_i\}, \quad C^{-1} = (p_{ij}). \quad (3.10)$$

This assumes, of course, that the matrix C is non-singular.

An interpretation similar to that for C_{ij} can be given for the elements p_{ij}, which are called coefficients of potential and which must depend only on the shapes and relative positions of the conductors, and not upon the charges or potentials. Suppose that all the conductors are uncharged except one, S_K, which carries unit (positive) charge. Then the potentials of the conductors are

$$p_{1K}, p_{2K}, \ldots, p_{KK}, \ldots, p_{NK}.$$

By taking $K = 1, 2, \ldots, N$ in succession this gives an interpretation for each element of the matrix (p_{ij}). Since only S_K carries a net charge we can draw certain conclusions. Tubes of force of aggregate strength one originate on S_K and these tubes must either end on infinity (the earth) or on one of the other conductors S_i. These two possibilities are illustrated in Fig. 3.4. Because all the other conductors S_i are uncharged, for every such tube which ends on S_i another of equal strength must originate on S_i. This tube must either go to infinity or end on another conductor S_j; the tube in question cannot end on S_K because potential decreases along the length of the tube and this tube is arising from a potential lower than V_K. For every such tube which ends on S_j another originates on S_j, and so on. We conclude that, since no tubes end on S_K, every point of S_K carries a positive surface density of charge and that every other conductor must have regions of positive and regions of negative surface density. Further, because the value of the potential decreases along every

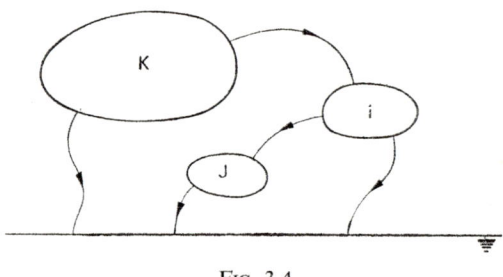

FIG. 3.4

line of force, we conclude that the potential of every conductor S_i lies between the potential of S_K and that of the earth, zero. It follows that

$$p_{KK} > p_{iK} \geqslant 0 \qquad (i \neq K, \; K = 1, 2, \ldots, N), \tag{3.11}$$

i.e. in each column of the matrix C^{-1} that element on the principal diagonal is greatest, and all elements are positive (or perhaps zero). This separation of negative and positive charges on each of the S_i when S_K alone is charged is the process of influence.

We prove first of all the important result that the matrix C (and its inverse C^{-1}) are symmetric, i.e. $c_{ij} = c_{ji}$, $p_{ij} = p_{ji}$. Consider two arbitrary distributions of charge, Q_i, Q'_i on the conductor S_i which correspond to the potential functions V, V', i.e.

V corresponds to potential V_i and charge Q_i on S_i,

V' corresponds to potential V'_i and charge Q'_i on S_i, ($i = 1, 2, \ldots, N$).

Then

$$Q_i = -\varepsilon_0 \oiint_{S_i} \frac{\partial V}{\partial \nu} \, dS, \quad Q'_i = -\varepsilon_0 \iint_{S_i} \frac{\partial V'}{\partial \nu} \, dS,$$

$$V = V_i \quad \text{on} \quad S_i, \qquad V' = V'_i \quad \text{on} \quad S_i,$$

and here the normal derivatives are directed into the field from the conduc-

§ 3.3   CONDUCTORS IN THE ELECTROSTATIC FIELD   83

tors and both V and V' satisfy the standard boundary conditions at infinity. We apply Green's theorem in the form

$$\iiint_T (U\nabla^2 V - V\nabla^2 U)\,d\tau = \oiint_S \left(U\frac{\partial V}{\partial n} - V\frac{\partial U}{\partial n}\right)dS,$$

where the differentiation $\partial/\partial n$ is taken out of the volume of integration. We use $V' = U$ and take the volume T to be that outside all the conducting surfaces S_i and inside a very large surface S_0 (a sphere) which encloses all S_i. Within this volume

$$\nabla^2 V = \nabla^2 V' = 0 \quad \text{(Laplace's equation)}$$

and the volume integral in Green's theorem vanishes so that

$$\oiint_{S_0}\left(V'\frac{\partial V}{\partial n} - V\frac{\partial V'}{\partial n}\right)dS = \sum_i \oiint_{S_i}\left(V'\frac{\partial V}{\partial n} - V\frac{\partial V'}{\partial n}\right)dS,$$

where on the right-hand side the differentiation $\partial/\partial n$ is now taken from the conductor *into* the field. The standard boundary conditions ensure that the integral over S_0 tends to zero as S_0 tends to infinity. Therefore

$$\sum_i \oiint_{S_i}\left(V'\frac{\partial V}{\partial n}dS - V\frac{\partial V'}{\partial n}\right)dS = 0.$$

But

$$\sum_i \oiint_{S_i} V'\frac{\partial V}{\partial n}\,dS = \sum_i V'_i \oiint_{S_i}\frac{\partial V}{\partial n}\,dS = -\frac{1}{\varepsilon_0}\sum_i V'_i Q_i,$$

$$\sum_i \oiint_{S_i} V\frac{\partial V'}{\partial n}\,dS = \sum_i V_i \oiint_{S_i}\frac{\partial V'}{\partial n}\,dS = -\frac{1}{\varepsilon_0}\sum_i V_i Q'_i.$$

Hence we obtain *Green's reciprocal theorem*

$$\sum_i V_i Q'_i = \sum_i V'_i Q_i. \tag{3.12}$$

To show the symmetry of the coefficients we choose V_i to correspond to a unit charge on S_K and no charge on the other conductors. Then V corresponds to

$$Q_i = 0, 0, \ldots, Q_K = 1, 0, \ldots, 0;$$
$$V_i = p_{1K}, p_{2K}, \ldots, p_{KK}, \ldots, p_{NK}.$$

Also V' corresponds to a unit charge on S_L and no charge on the other conductors, i.e. V' corresponds to

$$Q'_i = 0, 0, \ldots, Q_L = 1, \ldots, 0, 0; \quad V'_i = p_{1L}, p_{2L}, \ldots, p_{LL}, \ldots, p_{NL}.$$

Therefore $\sum_i Q_i V'_i = p_{KL}, \quad \sum_i Q'_i V_i = p_{LK}.$

Therefore, $\quad p_{KL} = p_{LK}.$ (3.13)

Similarly we can show that

$$c_{KL} = c_{LK}.$$ (3.14)

Condensers or Capacitors

The name condenser is being superseded by the name capacitor and refers to a system of two conductors (and the earth) so arranged that all the tubes of force arising from the first plate (conductor) when it is charged must end on the other plate (conductor), i.e. the second, earthed conductor screens

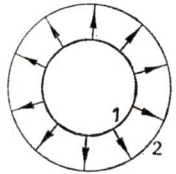
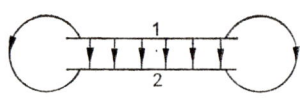

Fig. 3.5

the other. This is achieved in a spherical condenser (see p. 76) consisting of two concentric spheres when the charge is placed on the inner sphere. It is approximately achieved in the parallel plate condenser when the plates are plane conductors which are sufficiently close together for the "eakage" of the lines of force to be negligible, i.e. the field outside is negligilble (see Fig. 3.5). If this condition is achieved, the charges on the two plates must be equal and opposite. Here eqns. (3.6) become

$Q_1 = C_{11}V_1 + C_{12}V_2, \quad Q_2 = C_{12}V_1 + C_{22}V_2.$ with $Q_1 = q = -Q_2.$

When plate (2) is earthed and plate (1) is raised to unit potential

$$q = C_{11}, \quad -q = C_{12}.$$

Hence the coefficient of influence is equal to the first coefficient of capacitance. If the arrangement is reversible and all tubes of force arising from the second conductor must end on the first (this is not true for the spherical condenser but is approximately so for the parallel plate condenser) we can also put $V_1 = 0, V_2 = 1$ and obtain

$$-q' = C_{12}, \quad q' = C_{22}.$$

Hence we obtain

$$Q_1 = C_{11}(V_1 - V_2), \quad Q_2 = C_{11}(-V_1 + V_2) = -Q_1.$$

(We have used the symmetry of the coefficients earlier.)

Therefore,
$$C_{11} = \frac{Q_1}{V_1 - V_2}. \tag{3.15}$$

We see that C_{11} is the charge which, when placed on either plate, produces unit potential difference between them. This is called the *capacitance* of the condenser [cf. eqn. (3.5)].

Example 1. Find the capacity per unit area of a condenser consisting of two infinite conducting parallel planes whose distance apart is h.

An infinite uncharged sheet of metal of thickness a is introduced between the plates of such a condenser so that it is parallel to the plates. Show that the capacity per unit area of the condenser is increased by an amount $\varepsilon_0 a / \{h(h-a)\}$.

The field between the plates is
$$\mathbf{D} = \sigma \hat{\mathbf{v}}, \quad V = V_1 - (\sigma/\varepsilon_0)z, \tag{3.6}$$
where z is measured along the normal to conductor no. 1, Fig. 3.6 (i). Therefore
$$V_1 - V_3 = h\sigma/\varepsilon_0.$$
The charge on area α of each plate is
$$Q_1 = \sigma \alpha, \quad Q_2 = -\sigma \alpha.$$
Therefore,
$$\frac{Q_1}{V_1 - V_3} = \frac{\varepsilon_0 \sigma \alpha}{h \sigma} = \frac{\varepsilon_0 \alpha}{h}.$$
Therefore the capacitance of unit area is ε_0/h. This is an important result.

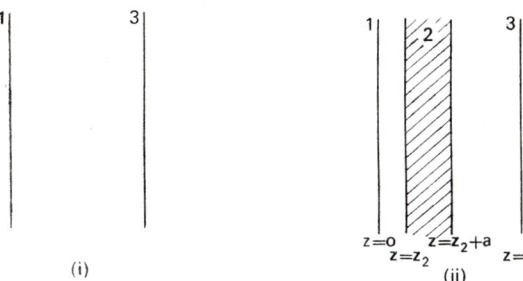

Fig. 3.6

When the uncharged sheet is introduced, Fig. 3.6 (ii), in the region between 1 and 2, i.e. for $0 \leq z \leq z_2$,
$$V = V_1 - (\sigma/\varepsilon_0)z.$$
In the region between 2 and 3, i.e. for $z_2 + a \leq z \leq h$,
$$V = V_1 - (\sigma/\varepsilon_0)z_2 - (\sigma/\varepsilon_0)(z - z_2 - a) = V_1 - (\sigma/\varepsilon_0)(z-a).$$
Therefore,
$$V_1 - V_3 = (\sigma/\varepsilon_0)(h-a).$$
Hence, the new capacitance is $\varepsilon_0/(h-a)$ so that the increase is
$$\frac{\varepsilon_0}{h-a} - \frac{\varepsilon_0}{h} = \frac{\varepsilon_0 a}{h(h-a)}.$$

Example 2. We give the complete set of equations for the spherical condenser.

$$Q_1 = C_{11}V_1 + C_{12}V_2, \quad Q_2 = C_{21}V_1 + C_{22}V_2. \tag{1}$$

When unit charge is placed upon the inner sphere (of radius a) the field both between the two spheres and outside the larger (of radius b) is that of a unit charge at the origin. Therefore $V = \dfrac{1}{4\pi\varepsilon_0 r}$; in particular $V_1 = \dfrac{1}{4\pi\varepsilon_0 a}$, $V_2 = \dfrac{1}{4\pi\varepsilon_0 b}$. When a unit charge is placed upon the outer sphere, and the inner sphere is left uncharged, there is no field inside $r = b$ and $V_1 = V_2$. Outside, the field is as before and so $V_2 = (4\pi\varepsilon_0 b)^{-1}$. Hence we obtain, by substituting these values into (1),

$$1 = C_{11}\frac{1}{4\pi\varepsilon_0 a} + C_{12}\frac{1}{4\pi\varepsilon_0 b}, \quad 0 = C_{21}\frac{1}{4\pi\varepsilon_0 a} + C_{22}\frac{1}{4\pi\varepsilon_0 b},$$

$$0 = C_{11}\frac{1}{4\pi\varepsilon_0 b} + C_{12}\frac{1}{4\pi\varepsilon_0 b}, \quad 1 = C_{21}\frac{1}{4\pi\varepsilon_0 b} + C_{22}\frac{1}{4\pi\varepsilon_0 b}.$$

Therefore $\quad C_{11} = -C_{12} = \dfrac{4\pi\varepsilon_0 ab}{b-a}, \quad C_{22} = \dfrac{4\pi\varepsilon_0 b^2}{b-a}, \quad C_{21} = -\dfrac{4\pi\varepsilon_0 ab}{b-a}.$

When the charge is placed on the inner sphere we have

$$Q_1 = \frac{4\pi\varepsilon_0 ab}{b-a}(V_1 - V_2),$$

so that the capacitance of the spherical condenser is $4\pi\varepsilon_0 ab/(b-a)$. The plates of this condenser are not interchangeable, as shown by the fact that $C_{22} \neq C_{11}$.

Example 3. We obtain the coefficients for the set of three concentric spheres of example 3, p. 77. Numbering the spheres 1, 2, 3 from the inside we first place

$$Q_1 = 1, \quad Q_2 = Q_3 = 0,$$

in which case $\quad V(r) = \dfrac{1}{4\pi\varepsilon_0 r}, \quad (r \geq a).$

Therefore $\quad V_1 = \dfrac{1}{4\pi\varepsilon_0 a}, \quad V_2 = \dfrac{1}{8\pi\varepsilon_0 a}, \quad V_3 = \dfrac{1}{12\pi\varepsilon_0 a}.$

With $Q_1 = 0$, $Q_2 = 1$, $Q_3 = 0$,

$$V(r) = V_2 \text{ for } r \leq 2a, \quad V(r) = \frac{1}{4\pi\varepsilon_0 r} \text{ for } r \geq 2a,$$

it follows that

$$V_1 = \frac{1}{8\pi\varepsilon_0 a} = V_2, \quad V_3 = \frac{1}{12\pi\varepsilon_0 a}.$$

Since with $Q_1 = 0$, $Q_2 = 0$, $Q_3 = 1$,

$$V_1 = V_2 = V_3 = V(r) \text{ for } r \leq 3a, \quad V(r) = \frac{1}{4\pi\varepsilon_0 r} \text{ for } r \geq 3a,$$

it follows that $\quad V_1 = V_2 = V_3 = \dfrac{1}{12\pi\varepsilon_0 a}.$

These when substituted into eqn. (3.7) give the three sets of equations

$$4\pi\varepsilon_0 a = C_{11} + \tfrac{1}{2}C_{12} + \tfrac{1}{3}C_{13}, \quad 0 = \tfrac{1}{2}(C_{11}+C_{12}) + \tfrac{1}{3}C_{13}, \quad 0 = \tfrac{1}{3}(C_{11}+C_{12}+C_{13}),$$

$$0 = C_{21} + \tfrac{1}{2}C_{22} + \tfrac{1}{3}C_{23}, \quad 4\pi\varepsilon_0 a = \tfrac{1}{2}(C_{21}+C_{22}) + \tfrac{1}{3}C_{23}, \quad 0 = \tfrac{1}{3}(C_{21}+C_{22}+C_{23}),$$

$$0 = C_{31} + \tfrac{1}{2}C_{32} + \tfrac{1}{3}C_{33}, \quad 0 = \tfrac{1}{2}(C_{31}+C_{32}) + \tfrac{1}{3}C_{33}, \quad 4\pi\varepsilon_0 a = \tfrac{1}{3}(C_{31}+C_{32}+C_{33}).$$

§ 3.3   CONDUCTORS IN THE ELECTROSTATIC FIELD

The solution of these equations corresponds to the equations
$$Q_1 = 4\pi\varepsilon_0 a(2V_1 - 2V_2), \quad Q_2 = 4\pi\varepsilon_0 a(-2V_1 + 8V_2 - 6V_3),$$
$$Q_3 = 4\pi\varepsilon_0 a(-6V_2 + 9V_3).$$

It can easily be verified that, on substituting the values of V_1, V_2, V_3 given in example 3, p. 77, we obtain
$$Q_1 = Q, \quad Q_2 = -2Q, \quad Q_3 = 3Q$$

which are the values given in that question.

Example 4. Two insulated spherical conductors of radii a and b are at a large distance r apart. One conductor carries a charge Q and the other is uncharged. Show that if the conductors are connected by a fine metallic wire, the charge on the conductor of radius a will become
$$\frac{Qa}{a+b}\left[1 + \frac{(a-b)b}{(a+b)r}\right], \quad \text{approximately.}$$

We regard this as a system of two conductors, and determine the coefficients of potential. Since the conductors are a large distance apart we can assume as a first approximation that, when one is charged and the other is uncharged, the charge is uniformly distributed on the charged sphere, and at large distance has the same effect as a point charge.

Hence,
(1) when $Q_1 = 1, Q_2 = 0$ we deduce that
$$V_1 = \frac{1}{4\pi\varepsilon_0 a}, \quad V_2 = \frac{1}{4\pi\varepsilon_0 r},$$

(2) when $Q_1 = 0, Q_2 = 1$,
$$V_1 = \frac{1}{4\pi\varepsilon_0 r}, \quad V_2 = \frac{1}{4\pi\varepsilon_0 b}.$$

But
$$V_1 = p_{11}Q_1 + p_{12}Q_2, \quad V_2 = p_{21}Q_1 + p_{22}Q_2. \tag{1}$$

Therefore
$$p_{11} = \frac{1}{4\pi\varepsilon_0 a}, \quad p_{12} = \frac{1}{4\pi\varepsilon_0 r} = p_{21}, \quad p_{22} = \frac{1}{4\pi\varepsilon_0 b}.$$

In the situation given in the question, we have, after the connection, in eqn. (1),
$$Q_1 + Q_2 = Q, \quad V_1 = V_2,$$

where
$$\frac{Q_1}{a} + \frac{Q_2}{r} = \frac{Q_1}{r} + \frac{Q_2}{b}. \tag{2}$$

Solving for Q_1 we obtain
$$\frac{Q_1}{Q} = \frac{1/b - 1/r}{1/a + 1/b - 2/r}.$$

In this expression $1/r$ is small, and the expression is only a first approximation, so we expand by the binomial theorem as far as the first power of $1/r$ as follows:
$$\frac{Q_1}{Q} = \left(\frac{1}{b} - \frac{1}{r}\right)\left(\frac{1}{a} + \frac{1}{b} - \frac{2}{r}\right)^{-1} = \frac{a}{a+b}\left(1 - \frac{b}{r}\right)\left\{1 - \frac{2ab}{(a+b)r}\right\}^{-1}$$
$$= \frac{a}{a+b}\left\{1 - \frac{b}{r} + \frac{2ab}{(a+b)r}\right\} = \frac{a}{a+b}\left\{1 + \frac{b(a-b)}{(a+b)r}\right\}.$$

Example 5. The cylindrical capacitor.

This arrangement consists of two long coaxial cylindrical conductors of radii a and b ($> a$), Fig. 3.7, and is the two-dimensional counterpart of the spherical capacitor.

Using cylindrical polar coordinates, Laplace's equation is

$$\frac{\partial^2 V}{\partial r^2} + \frac{1}{r}\frac{\partial V}{\partial r} + \frac{1}{r^2}\frac{\partial^2 V}{\partial \theta^2} + \frac{\partial^2 V}{\partial z^2} = 0.$$

For the region between the cylinders, $a \leqslant r \leqslant b$, the potential function is independent of θ, z and depends on r only. Therefore

$$\frac{d^2 V}{dr^2} + \frac{1}{r}\frac{dV}{dr} = 0, \quad V = A + B \ln r$$

[cf. eqn. (2.18)]. When the outer conductor is earthed, $V = V_1$ where $r = a$, $V = 0$ where $r = b$. Therefore

$$V_1 = A + B \ln a, \quad 0 = A + B \ln b, \quad V_1 = B \ln (a/b). \tag{1}$$

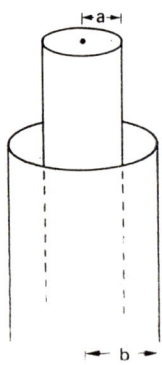

FIG. 3.7

The electric field is $E = -\text{grad } V$, therefore

$$E_r = -\frac{dV}{dr} = -\frac{B}{r}, \quad E_\theta = E_z = 0.$$

Near $r = a$, where the (surface) charge density is σ, we have

$$D_r = \sigma, \quad D_\theta = D_z = 0.$$

Hence $\sigma = -\varepsilon_0 B/a$ and it follows that $V_1 = -(\sigma/a\varepsilon_0) \ln (a/b)$. The charge on unit length of the inner cylinder is $e = 2\pi a \sigma$. Therefore

$$V_1 = -\frac{e}{2\pi\varepsilon_0} \ln\left(\frac{a}{b}\right),$$

and the capacitance of unit length of the conductors is

$$C = \frac{e}{V_1} = \frac{2\pi\varepsilon_0}{\ln(b/a)}. \tag{2}$$

The capacitance of the whole length of the cylinders is infinite so, for two-dimensional fields of this nature, the capacitance is always given for unit length. However, the formula (2) applies *only* if the field is strictly two-dimensional, i.e. when the actual length of each cylinder is large compared with a or b.

§ 3.3 CONDUCTORS IN THE ELECTROSTATIC FIELD

Example 6. Condensers in series and parallel.

(a) *Series*. Because all the tubes of force arising from one plate of a condenser must end on the other, when C_1 receives a charge q as shown in Fig. 3.8 (i) there must also be an arrangement of charges $\pm q$ on the plates of the other condensers as shown. (B need not be

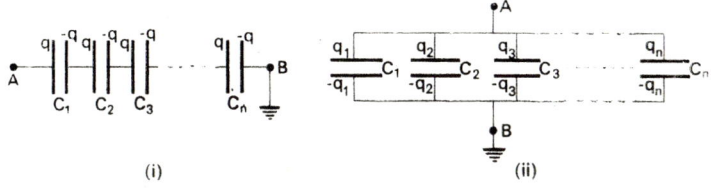

FIG. 3.8

connected to earth but often is.) The potential difference between A and B is therefore the sum of the potential differences of the various condensers and so

$$V_A - V_B = \frac{q}{C_1} + \frac{q}{C_2} + \ldots + \frac{q}{C_n}.$$

A single condenser with one plate connected to A and the other to B receiving a charge q would have a capacity C if the potential difference between its plates were $V_A - V_B$ where $C = q/(V_A - V_B)$. It follows that

$$\frac{1}{C} = \frac{1}{C_1} + \frac{1}{C_2} + \ldots + \frac{1}{C_n}.$$

(b) *Parallel*. Since all the upper plates are connected to A and the lower plates to B [see Fig. 3.8 (ii)] the charges must arrange themselves on each plate so that

$$V_A - V_B = \frac{q_1}{C_1} = \frac{q_2}{C_2} = \frac{q_3}{C_3} = \ldots = \frac{q_n}{C_n}.$$

Therefore a total charge $q = q_1 + q_2 + \ldots + q_n$ must be placed at A in order to give this distribution. Here a single condenser connected between A and B receiving a charge q would have a capacitance C where

$$C = \frac{q}{V_A - V_B} = \frac{q_1 + q_2 + \ldots + q_n}{V_A - V_B}.$$

Hence

$$C = C_1 + C_2 + \ldots + C_n.$$

In each of the above cases the capacitance C is called the "equivalent capacitance" for the given arrangement of condensers.

Exercises 3.3

1. Calculate the capacity of a system consisting of two concentric spheres S_a, S_b with radii a, b ($b > a$) and deduce the capacity of an isolated sphere.

S_a and S_b are surrounded by a third concentric sphere S_c of radius c which is connected with S_a by a fine wire which is insulated from S_b. Find the capacity of the condenser so formed.

2. Three insulated concentric spherical conductors, whose radii (in ascending order of magnitude) are a, b, c, have charges e_1, e_2, e_3 respectively. Find their potentials, and show that, if the innermost sphere be connected to earth, the potential of the outermost is diminished by

$$\frac{a}{4\pi\varepsilon_0 c}\left(\frac{e_1}{a}+\frac{e_2}{b}+\frac{e_3}{c}\right).$$

3. A spherical conductor of radius a and carrying a charge e_1 is surrounded by a concentric spherical conducting sheet of radius b and carrying a charge e_2, both conductors being insulated. Find the potential at a point between the spheres.

If the inner conductor is connected by a fine insulated conducting wire, passing through a small hole in the outer conductor, to a distant uncharged, insulated spherical conductor of radius c, prove that the latter will be raised to a potential

$$\frac{e_1 b + e_2 a}{4\pi\varepsilon_0 b(a+c)}.$$

4. If two surfaces be taken in any family of equipotentials in free space, and two metal conductors formed so as to occupy their positions, then the capacity of the condenser thus formed is

$$\frac{C_1 C_2}{C_1 - C_2},$$

where C_1, C_2 are the capacities of the external and internal conductors when existing alone in an infinite field.

3.4 Forces and energy

We turn now to an aspect of the theory which has not so far been considered, namely that of the energy required to establish an electrostatic field. In the picture of a stretched string, used as an illustration in Chapter 1, when energy has been stored in the string a force is exerted by the string at either end. For the special case of a string obeying Hooke's law the energy is given by $W = \lambda x^2/(2a)$ when the displacement at the end is x. Furthermore, the force exerted by the string which tends to *increase* x is the tension, given by

$$T = -\frac{\lambda x}{a} = -\frac{dW}{dx}.$$

(This result is characteristic of the general relation between force and potential energy in mechanics.)

We have already established two fundamental results concerning the forces experienced by charges in an electrostatic field. They are:

(1) a point charge Q, at a place where the field caused by all charges excluding Q itself is E, experiences a force QE;

(2) a small area α of surface carrying charge density σ, where the fields on the sides of α are E_+, E_-, experiences a force $\frac{1}{2}\alpha\sigma(E_+ + E_-)$; this force arises from the field caused by all charges except the immediate local surface charge on α, i.e. the force per unit area on such a surface distribution is $\frac{1}{2}\sigma(E_+ + E_-)$.

In the case of a surface charge on a conductor this force is always equivalent to an outward force per unit area of $\sigma^2/(2\varepsilon_0)$.

Also we can expect a force to act on a small volume v containing a charge density ϱ. Since the contribution to the field from v itself becomes vanishingly small as v is reduced we expect that a force ϱE per unit volume acts at a point of a volume charge distribution where the electric field is E. The action on any finite distribution of charge could be obtained by combining these forces using integrations where necessary to give a total resultant force at any point together with a couple.

Our aim in this discussion of energy and its relation to the forces which act upon bodies and charges may be roughly stated as follows. The charges which are the sources of an electrostatic field are regarded as having been assembled in their final positions from dispersal at infinity. In assembling the charges work was done against the mutual interactions of the charges, and this work is regarded as *the potential energy of the field*. If the configuration of the field alters, the forces acting on any body, or charge, in the field do work, and this work is done at the expense of the potential energy. The configuration of any field, i.e. the positions and orientations of conductors, the positions of point charges, etc., may be specified by a number of generalized coordinates (see *C.M.*, vol. 3, chap. 2; vol. 6, chap. 2) of which we take x to be typical. Corresponding to each of these coordinates is a generalized force X which tends to increase x. (The external agents holding the whole system stationary have to exert a force $-X$ to balance each force X exerted by the field.) The forces exerted by the field are often called ponderomotive forces. In a change of configuration in which each x increases by δx the ponderomotive forces do work $\sum X\,\delta x$; this work is done at the expense of the potential energy W. Therefore,

$$-\delta W = \sum X\,\delta x. \tag{3.16}$$

When the different coordinates x are independent we deduce that

$$X = -\frac{\partial W}{\partial x}. \tag{3.17}$$

We consider now a number of important special cases.

1. A POINT CHARGE IN A FIELD

The work necessary to place a charge Q at a point P is

$$W = QV, \tag{3.18}$$

where V is the potential of the field at P. This follows from the definition of V. The cartesian coordinates of the point charge are $\{x\ y\ z\}$ so that the force on the charge has resolutes

$$X = -\frac{\partial W}{\partial x} = -Q\frac{\partial V}{\partial x} = QE_x, \quad Y = QE_y, \quad Z = QE_z,$$

i.e. $\qquad\qquad F = -\text{grad } W = -Q \text{ grad } V = QE.$

Here we have assumed that the field is unaffected by the charge Q and by the displacement of Q. We may call this field a *rigid* field.

2. ACTION ON A DIPOLE OF MOMENT μ

The potential energies of the two charges which in the limit constitute a dipole are $-QV$, $Q(V+\delta V)$, where $\delta V = \boldsymbol{l}\cdot\text{grad } V$. Therefore

$$\begin{aligned} W &= \lim \{-QV+QV+Q\boldsymbol{l}\cdot\text{grad } V\} \\ &= \boldsymbol{\mu}\cdot\text{grad } V = -\boldsymbol{\mu}\cdot\boldsymbol{E}. \end{aligned} \tag{3.19}$$

The most general small displacement of a dipole consists of a rigid body displacement $(\boldsymbol{\delta s}, \boldsymbol{\delta\theta})$ made up of a translation $\boldsymbol{\delta s}$, and an infinitesimal rotation $\boldsymbol{\delta\theta}$. The moment μ is unaltered by $\boldsymbol{\delta s}$ but is altered by the rotation. Therefore,

$$\delta\boldsymbol{\mu} = \boldsymbol{\delta\theta}\times\boldsymbol{\mu}.$$

Hence, in such a displacement,

$$\begin{aligned} \delta W &= -(\boldsymbol{\mu}+\delta\boldsymbol{\mu})\cdot(\boldsymbol{E}+\delta\boldsymbol{E})+(\boldsymbol{\mu}\cdot\boldsymbol{E}) = -\boldsymbol{\mu}\cdot\delta\boldsymbol{E}-\delta\boldsymbol{\mu}\cdot\boldsymbol{E} \\ &= -\boldsymbol{\mu}\cdot\delta\boldsymbol{E}-(\boldsymbol{\delta\theta}\times\boldsymbol{\mu})\cdot\boldsymbol{E} = -\boldsymbol{\mu}\cdot\delta\boldsymbol{E}-(\boldsymbol{\mu}\times\boldsymbol{E})\cdot(\boldsymbol{\delta\theta}). \end{aligned}$$

The general action on a dipole consists of a force $\boldsymbol{F}$ and a couple $\boldsymbol{\Gamma}$ at P. The work done in an arbitrary displacement by these is $\boldsymbol{F}\cdot\boldsymbol{\delta s}+\boldsymbol{\Gamma}\cdot\boldsymbol{\delta\theta}$. The conservation of energy requires that

$$\boldsymbol{F}\cdot\boldsymbol{\delta s}+\boldsymbol{\Gamma}\cdot\boldsymbol{\delta\theta} = -\delta W = \boldsymbol{\mu}\cdot\delta\boldsymbol{E}+\boldsymbol{\delta\theta}\cdot(\boldsymbol{\mu}\times\boldsymbol{E}).$$

From this, since $\boldsymbol{\delta\theta}$ is arbitrary, we deduce

$$\boldsymbol{\Gamma} = \boldsymbol{\mu}\times\boldsymbol{E}. \tag{3.20}$$

Also, since $\qquad\qquad \delta\boldsymbol{E} = (\boldsymbol{\delta s}\cdot\nabla)\boldsymbol{E}.$

$$\boldsymbol{F}\cdot\boldsymbol{\delta s} = \boldsymbol{\mu}\cdot[(\boldsymbol{\delta s}\cdot\nabla)\boldsymbol{E}].$$

§ 3.4   CONDUCTORS IN THE ELECTROSTATIC FIELD   93

Since μ is a constant, $\mu \cdot [(\delta s \cdot \nabla)E] = (\delta s \cdot \nabla)(\mu \cdot E)$.
Therefore, $F = \nabla(\mu \cdot E) = -\mathbf{grad}\ W$,
which agrees with the general result above. Since μ is a constant vector (i.e. independent of the variables of differentiation in ∇), $\nabla(\mu \cdot E) = \mu \times \mathbf{curl}\ E + (\mu \cdot \nabla)E$. Because $\mathbf{curl}\ E = 0$ we obtain

$$F = (\mu \cdot \nabla)E, \tag{3.21}$$

the result given in eqns. (2.38) and (2.39).

Example. A dipole M_2 can rotate freely about its centre P. Deduce that if another dipole M_1 is fixed at Q, the dipole M_2 is in equilibrium when it lies in the same plane as M_1 and $r = \overrightarrow{QP}$ and

$$\tan \theta_1 + 2 \tan \theta_2 = 0,$$

θ_1, θ_2 being the angles between r and M_1, M_2 respectively.

Show that, if both dipoles can rotate freely about their centres, they must lie along, or perpendicular to, the direction of r if they are to be in equilibrium.

Using eqn. (3.19) the potential energy of M_2 in the field of M_1 is

$$W = + \frac{M_1 \cdot M_2}{r^3} - \frac{3(M_1 \cdot r)(M_2 \cdot r)}{r^5}.$$

The only displacement possible for M_2 is a rotation about its centre. If we give M_2 an infinitesimal rotation $\boldsymbol{\delta\phi}$ (see *C.M.*, vol. 6, § 1.6), $\boldsymbol{\delta M_2} = \boldsymbol{\delta\phi} \times M_2$. Therefore

$$\delta W = \frac{M_1 \cdot (\boldsymbol{\delta\phi} \times M_2)}{r^3} - \frac{3(M_1 \cdot r)[(\boldsymbol{\delta\phi} \times M_2) \cdot r]}{r^5} = \left\{ \frac{M_2 \times M_1}{r^3} - \frac{3(M_1 \cdot r)(M_2 \times r)}{r^5} \right\} \cdot \boldsymbol{\delta\phi}.$$

By the principle of virtual work $\delta W = 0$ when the system is in equilibrium, $\boldsymbol{\delta\phi}$ being an arbitrary vector. Therefore

$$0 = \left\{ \frac{M_2 \times M_1}{r^3} - \frac{3(M_1 \cdot r)(M_2 \times r)}{r^5} \right\} \cdot \boldsymbol{\delta\phi}.$$

Hence
$$(M_2 \times M_1)r^2 = 3(M_1 \cdot r)(M_2 \times r). \tag{1}$$

Taking the scalar product of eqn. (1) with r we obtain

$$r^2(M_2 \times M_1) \cdot r = 3(M_1 \cdot r)[(M_2 \times r) \cdot r] = 0.$$

Since $r \neq 0$ we deduce that $(M_2 \times M_1) \cdot r = 0$, i.e. M_1, M_2 and r are coplanar vectors.

Suppose now that M_1 and M_2 make angles θ_1, θ_2 in this plane with the direction of r, both angles being measured in the same sense. Then we equate the magnitude of each side of eqn. (1), each being a vector perpendicular to the plane, obtaining

$$M_1 M_2 r^2 \sin(\theta_2 - \theta_1) = 3(M_1 r \cos \theta_1)(M_2 r \sin \theta_2);$$
i.e. $$\sin(\theta_2 - \theta_1) = 3 \cos \theta_1 \sin \theta_2$$

which leads to $2 \tan \theta_2 + \tan \theta_1 = 0$.

When both M_1 and M_2 can rotate freely about their centres we apply a similar method in which $\boldsymbol{\delta M_1} = \boldsymbol{\delta\phi}_1 \times M_1$, $\boldsymbol{\delta M_2} = \boldsymbol{\delta\phi}_2 \times M_2$. Then, in a small displacement from the equilibrium position defined by rotations $\boldsymbol{\delta\phi}_1, \boldsymbol{\delta\phi}_2$,

$$\delta W = 0 = + \frac{\boldsymbol{\delta\phi}_1 \cdot (M_1 \times M_2) - \boldsymbol{\delta\phi}_2 \cdot (M_2 \times M_1)}{r^3}$$

$$- \frac{3\boldsymbol{\delta\phi}_1 \cdot (M_1 \times r)(M_2 \cdot r) + 3(M_1 \cdot r)\boldsymbol{\delta\phi}_2 \cdot (M_2 \times r)}{r^5}.$$

Since both $\boldsymbol{\delta\phi}_1$ and $\boldsymbol{\delta\phi}_2$ are arbitrary we deduce that

$$\frac{\mathbf{M}_1 \times \mathbf{M}_2}{r^3} - \frac{3(\mathbf{M}_1 \times \mathbf{r})(\mathbf{M}_2 \cdot \mathbf{r})}{r^5} = \mathbf{0}, \quad \frac{\mathbf{M}_2 \times \mathbf{M}_1}{r^3} + \frac{3(\mathbf{M}_1 \cdot \mathbf{r})(\mathbf{M}_2 \times \mathbf{r})}{r^5} = \mathbf{0}. \quad (2)$$

On taking the scalar product of each of these by $\mathbf{r}$ we see that

$$(\mathbf{M}_1 \times \mathbf{M}_2) \cdot \mathbf{r} = 0,$$

i.e. that $\mathbf{M}_1, \mathbf{M}_2, \mathbf{r}$ are coplanar, as before. Using the angles θ_1, θ_2 introduced above we now obtain, from the magnitude of the vectors in eqn. (2), that

$$M_1 M_2 \sin(\theta_2 - \theta_1) - 3 M_1 M_2 \sin\theta_1 \cos\theta_2 = 0,$$
$$M_1 M_2 \sin(\theta_2 - \theta_1) + 3 M_1 M_2 \cos\theta_1 \sin\theta_2 = 0.$$

Therefore $\quad \sin\theta_1 \cos\theta_2 + \cos\theta_1 \sin\theta_2 = \sin(\theta_1 + \theta_2) = 0,$

i.e. $\quad \theta_1 + \theta_2 = n\pi \quad (n = 0, 1, 2, \ldots).$

Therefore $\sin(n\pi - 2\theta_1) - 3\sin\theta_1 \cos(n\pi - \theta_1) = (-1)^{n-1} \sin 2\theta_1 + 3(-1)^n \sin\theta_1 \cos\theta_1 = 0.$

Therefore $\quad -2\sin\theta_1 \cos\theta_1 + 3\sin\theta_1 \cos\theta_1 = \sin\theta_1 \cos\theta_1 = 0,$

i.e. $\quad \sin\theta_1 = 0 \quad \text{or} \quad \cos\theta_1 = 0.$

Hence the equilibrium positions occur when $\mathbf{M}_1, \mathbf{M}_2$ are both directed along $\mathbf{r}$, either parallel or anti-parallel, or when $\mathbf{M}_1$ and $\mathbf{M}_2$ are both at right-angles to $\mathbf{r}$, again either parallel or anti-parallel to each other.

In the above example we considered the work necessary in order to establish a charge, or dipole, in an existing *rigid* field. Now we consider the case in which a field is built up element by element from nothing and so derive an expression for the work that must be done to build up the field. We call the energy found above the *mutual (potential) energy* of the existing rigid system and the new system; we call the energy we determine now the *(self-) energy* of the system.

(1) We consider an isolated collection of point charges Q_l ($l = 1, 2, \ldots, N$) assembled in order from infinite dispersion. To place the charge Q_1 in its final position requires no work; to place charge Q_2 in position requires work $Q_1 Q_2/(4\pi\varepsilon_0 r_{12})$ to be done against the field of charge Q_1; to bring Q_3 into position requires work

$$\frac{Q_1 Q_3}{4\pi\varepsilon_0 r_{13}} + \frac{Q_2 Q_3}{4\pi\varepsilon_0 r_{23}}$$

to be done and so on. (Here r_{lm} is the distance between Q_l and Q_m.) The total work required to place all N charges in position is

$$W = \frac{Q_1 Q_2}{4\pi\varepsilon_0 r_{12}} + \left(\frac{Q_1 Q_3}{4\pi\varepsilon_0 r_{13}} + \frac{Q_2 Q_3}{4\pi\varepsilon_0 r_{23}}\right) + \left(\frac{Q_1 Q_4}{4\pi\varepsilon_0 r_{14}} + \frac{Q_2 Q_4}{4\pi\varepsilon_0 r_{24}} + \frac{Q_3 Q_4}{4\pi\varepsilon_0 r_{34}}\right) + \ldots$$

$$= \frac{1}{4\pi\varepsilon_0} \sum_{m > l} \frac{Q_l Q_m}{r_{lm}} = \frac{1}{4\pi\varepsilon_0} \frac{1}{2} \sum_{l \neq m} \frac{Q_l Q_m}{r_{lm}}.$$

§ 3.4   CONDUCTORS IN THE ELECTROSTATIC FIELD

Therefore
$$W = \frac{1}{8\pi\varepsilon_0} \sum \frac{Q_l Q_m}{r_{lm}}. \tag{3.22}$$

The coefficient of Q_l in the last expression is $\frac{1}{2} \sum_{m \neq l} \frac{Q_m}{4\pi\varepsilon_0 r_{ml}}$. This is half the potential V_l produced at the position of charge Q_l by all the remaining charges. Therefore

$$W = \frac{1}{2} \sum_{l=1}^{N} Q_l V_l. \tag{3.23}$$

(2) We consider a field arising from a number of charged conductors S_i. If the final charges and potentials are Q_i, V_i, the linear relations (3.6) show that when each conductor carries a fraction λ of its final charge, viz. λQ_i ($0 \leq \lambda \leq 1$), the potential of each conductor is λV_i. We arrange to increase the charges on the conductors to their final values by adding an amount $Q_i d\lambda$ simultaneously to the respective conductors. To add this extra charge to the conductors when at the potential λV_i requires work

$$dW = \sum_i (\lambda V_i)(d\lambda\, Q_i) + O\{(d\lambda)^2\}.$$

Therefore
$$W = \sum V_i Q_i \int_0^1 \lambda\, d\lambda = \frac{1}{2} \sum_i V_i Q_i. \tag{3.24}$$

(3) We consider a field arising from a *rigid* distribution of volume and surface charge ρ, σ. Again, we build up the charges as in the case of the conductors by increasing ρ, σ from zero to their final values in proportion to these final values. To do this we add a charge $(\rho\, d\tau)d\lambda$ to a typical volume element $d\tau$ when there already exists a charge distribution $\lambda\rho$, $\lambda\sigma$. The surface charges increase similarly. The potential at the element $d\tau$ in the intermediate stage is λV, where V is the final potential at this element. Hence the work done at a typical element $d\tau$ or dS is

$$(\rho V d\tau)\lambda\, d\lambda \quad \text{or} \quad (\sigma V dS)\lambda\, d\lambda.$$

Hence
$$dW = \left[\iiint \rho V\, d\tau + \iint \sigma V\, dS \right] \lambda\, d\lambda$$

and
$$W = \left[\iiint \rho V\, d\tau + \iint \sigma V\, dS \right] \int_0^1 \lambda\, d\lambda = \frac{1}{2} \iiint \rho V\, d\tau + \frac{1}{2} \iint \sigma V\, dS. \tag{3.25}$$

Example 1. For the field of a number of conductors we can express W in two other forms.
$$W = \tfrac{1}{2} \sum V_i Q_i = \tfrac{1}{2} \sum_{ij} p_{ij} V_i V_j = \tfrac{1}{2} \sum_{ij} C_{ij} Q_i Q_j.$$

In the special case of an isolated conductor of capacitance C or of a pair of conductors forming a condenser this becomes

$$W = \tfrac{1}{2}VQ = \tfrac{1}{2}CV^2 = \tfrac{1}{2}Q^2/C.$$

Example 2. A system of conductors carrying charges Q_i at potentials V_i has energy W, and the same system of conductors carrying charges Q'_i at potentials V'_i has energy W'. Then

$$W' - W = \tfrac{1}{2}\sum_i (Q'_i V'_i - Q_i V_i).$$

Now

$$\sum_i (Q'_i - Q_i)(V'_i - V_i) = \sum_i (Q'_i V'_i - Q_i V_i) + \sum_i Q_i V_i - \sum_i Q'_i V'_i.$$

The last two terms cancel by Green's reciprocal theorem. Therefore

$$W' - W = \tfrac{1}{2}\sum_i (Q'_i - Q_i)(V'_i - V_i).$$

Similarly

$$W' - W = \tfrac{1}{2}\sum_i (Q'_i + Q_i)(V'_i - V_i).$$

These expressions are useful for calculating changes of energy in a system.

Example 3. For an isolated spherical shell of radius a carrying a charge Q,

$$V_1 = \frac{Q}{4\pi\varepsilon_0 a} \quad \text{and} \quad W = \frac{1}{2}QV_1 = \frac{Q^2}{8\pi\varepsilon_0 a}.$$

Example 4. For a thick shell of inner and outer radii b, a, carrying a charge Q distributed with a uniform density ϱ, we have (using the results of Example 5, p. 44)

$$\varrho = \frac{3Q}{4\pi(a^3 - b^3)}, \quad V(r) = \frac{\varrho}{6\varepsilon_0}\left(3a^2 - r^2 - \frac{2b^3}{r}\right) \quad (b \leqslant r \leqslant a).$$

Therefore
$$W = \frac{1}{2}\iiint \varrho V \, d\tau = \frac{1}{2}\int_b^a \frac{\varrho^2}{6\varepsilon_0}\left(3a^2 - r^2 - \frac{2b^3}{r}\right) 4\pi r^2 \, dr$$

$$= \frac{\pi\varrho^2}{3\varepsilon_0}\left\{a^2(a^3 - b^3) - \tfrac{1}{5}(a^5 - b^5) - b^3(a^2 - b^2)\right\}$$

$$= \frac{3Q^2}{16\pi\varepsilon_0(a^3 - b^3)^2}\left\{a^2(a^3 - b^3) - \tfrac{1}{5}(a^5 - b^5) - b^3(a^2 - b^2)\right\}.$$

In the special case of a uniform spherical distribution (with no cavity), $b \equiv 0$ and

$$W = \frac{3Q^2}{20\,\pi\varepsilon_0 a}.$$

Example 5. We consider two non-overlapping rigid spherical distributions S_a, S_b of uniform charge, denoted by the suffixes a, b, where $AB = c$, Fig. 3.9.

§ 3.4 CONDUCTORS IN THE ELECTROSTATIC FIELD

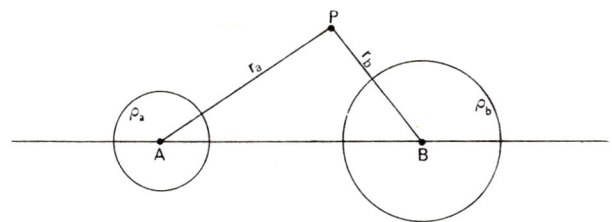

FIG. 3.9

The potential at P can be written as the sum of two contributions, $V = V_a + V_b$, where

$$V_a \begin{cases} = \dfrac{\varrho_a}{6\varepsilon_0}(3a^2 - r_a^2) & \text{for } 0 \leqslant r \leqslant a, \\ = \dfrac{\varrho_a a^3}{3\varepsilon_0 r_a} & \text{for } a \leqslant r_a; \end{cases}$$

$$V_b \begin{cases} = \dfrac{\varrho_b}{6\varepsilon_0}(3b^2 - r_b^2) & \text{for } 0 \leqslant r_b \leqslant b, \\ = \dfrac{\varrho_b b^3}{3\varepsilon_0 r_b} & \text{for } b \leqslant r_b. \end{cases}$$

From formula (3.25),

$$W = \frac{1}{2}\iiint_{S_a} V\varrho_a \, d\tau + \frac{1}{2}\iiint_{S_b} V\varrho_b \, d\tau$$

$$= \frac{1}{2}\iiint_{S_a} V_a\varrho_a \, d\tau + \frac{1}{2}\iiint_{S_a} V_b\varrho_a \, d\tau + \frac{1}{2}\iiint_{S_b} V_a\varrho_b \, d\tau + \frac{1}{2}\iiint_{S_b} V_b\varrho_b \, d\tau.$$

Now

$$\iiint_{S_a} V_b\varrho_a \, d\tau = \iiint_{S_a} \frac{\varrho_a \varrho_b b^3 \, d\tau}{3\varepsilon_0 r_b} = \varrho_b b^3 \iiint_{S_a} \frac{\varrho_a \, d\tau}{3\varepsilon_0 r_b}$$

$$= \frac{4\pi\varrho_b b^3}{3} \iiint_{S_a} \frac{\varrho_a \, d\tau}{4\pi\varepsilon_0 r_b} = Q_b V_a(B) = \frac{Q_a Q_b}{4\pi\varepsilon_0 c}.$$

Similarly

$$\iiint_{S_b} V_a\varrho_b \, d\tau = \frac{Q_a Q_b}{4\pi\varepsilon_0 c} \left[= \iiint_{S_a} V_b\varrho_a \, d\tau \right].$$

Also $\dfrac{1}{2}\iiint_{S_a} V_a\varrho_a \, d\tau = W_{aa}$, $\dfrac{1}{2}\iiint_{S_b} V_b\varrho_b \, d\tau = W_{bb}$ are the energies of S_a and S_b respectively in the absence of each other. Therefore

$$W = W_{aa} + W_{ab} + W_{bb},$$

where W_{aa}, W_{bb} are independent of c and

$$W_{ab} = \frac{Q_a Q_b}{4\pi\varepsilon_0 c}.$$

Hence the force between these distributions is

$$-\frac{\partial W}{\partial c} = \frac{Q_a Q_b}{4\pi\varepsilon_0 c^2}$$

tending to increase c. The spherical distributions behave as point charges.

We leave it an exercise for the reader to show that this result holds when the distributions of charge are each spherically symmetric but not uniform, i.e. ϱ_a is a function of r_a and ϱ_b is a function of r_b.

Example 6. Three concentric thin spherical conductors of radii a, b, c ($a < b < c$) have charges Q_1, Q_2, Q_3 respectively. Find their potentials and show that, if the two outer shells are connected by a thin wire, the loss of electrical energy is

$$\frac{1}{8\pi\varepsilon_0}\left(\frac{1}{b}-\frac{1}{c}\right)(Q_1+Q_2)^2.$$

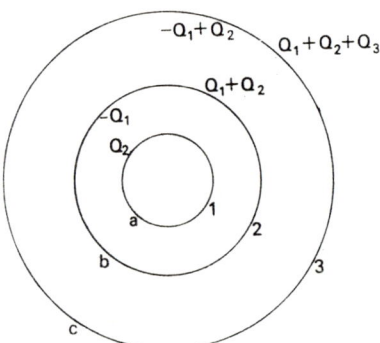

Fig. 3.10

To find the coefficients of potential of eqn. (3.6) in this case we write down the potentials when unit charges are placed, in succession, on the insulated conductors 3, 2, 1, the other conductors being uncharged, see Fig. 3.10.

When $Q_3 = 1$, $Q_2 = Q_1 = 0$, conductor 3 is an equipotential containing no charge. Hence the potential inside is the same as that of the conductor itself so that

$$Q_3 = 1; \quad V_3 = V_2 = V_1 = 1/(4\pi\varepsilon_0 c). \tag{1}$$

When $Q_2 = 1$, $Q_3 = Q_1 = 0$, conductor 2 contains no charge so that $V_2 = V_1 = 1/(4\pi\varepsilon_0 b)$. Effectively the field outside conductor 2 is that of a point charge so that $V_3 = 1/(4\pi\varepsilon_0 c)$ and

$$Q_2 = 1, \quad V_2 = V_1 = 1/(4\pi\varepsilon_0 b), \quad V_3 = 1/(4\pi\varepsilon_0 c). \tag{2}$$

When $Q_1 = 1$, $Q_2 = Q_3 = 0$, outside the first conductor the field is that of a point charge. In this case

$$Q_1 = 1, \quad V_1 = 1/(4\pi\varepsilon_0 a), \quad V_2 = 1/(4\pi\varepsilon_0 b), \quad V_3 = 1/(4\pi\varepsilon_0 c). \tag{3}$$

These determine the coefficients of potential in eqns. (3.9) which we write

$$4\pi\varepsilon_0 V_1 = Q_1/a + Q_2/b + Q_3/c, \quad 4\pi\varepsilon_0 V_2 = (Q_1+Q_2)/b + Q_3/c, \tag{4, 5}$$

$$4\pi\varepsilon_0 V_3 = (Q_1+Q_2+Q_3)/c. \tag{6}$$

These give the initial potentials of the conductors.

§ 3.4 CONDUCTORS IN THE ELECTROSTATIC FIELD

When conductors 2 and 3 are connected by a wire the charges on them are redistributed so that the two conductors take up the same potential. The new charges and potentials are:

$$Q_1' = Q_1, \quad Q_2', \quad Q_3' \text{ where } Q_2' + Q_3' = Q_2 + Q_3;$$
$$V_1', \quad V_2' \,(= V_3'), \quad V_3' \,(= V_2').$$

Utilizing eqns. (5, 6) with the redistributed charges we find

$$4\pi\varepsilon_0 V_2' = (Q_1 + Q_2')/b + Q_3'/c = 4\pi\varepsilon_0 V_3' = (Q_1 + Q_2' + Q_3')/c = (Q_1 + Q_2 + Q_3)/c.$$

Then
$$V_3' = V_3, \quad Q_1\left(\frac{1}{b}-\frac{1}{c}\right) + Q_2'\left(\frac{1}{b}-\frac{1}{c}\right) = 0, \quad Q_2' = -Q_1.$$

Therefore
$$Q_3' = Q_1 + Q_2 + Q_3.$$

Hence
$$4\pi\varepsilon_0 V_1' = Q_1/a - Q_1/b + (Q_1 + Q_2 + Q_3)/c,$$
$$4\pi\varepsilon_0 V_2' = 4\pi\varepsilon_0 V_3' = (Q_1 + Q_2 + Q_3)/c.$$

Using the result of example 2 on p. 96 the loss of electrical energy is

$$\tfrac{1}{2}\{(Q_1-Q_1')(V_1+V_1')+(Q_2-Q_2')(V_2+V_2')+(Q_3-Q_3')(V_3+V_3')\}$$

$$= \frac{1}{8\pi\varepsilon_0}\left\{(Q_2+Q_1)\left(\frac{Q_1+Q_2}{b}+\frac{Q_3}{c}+\frac{Q_1+Q_2+Q_3}{c}\right)\right.$$

$$\left. -(Q_1+Q_2)\left(\frac{Q_1+Q_2+Q_3}{c}+\frac{Q_1+Q_2+Q_3}{c}\right)\right\} = \frac{1}{8\pi\varepsilon_0}(Q_2+Q_1)^2\left(\frac{1}{b}-\frac{1}{c}\right).$$

Example 7. A spherical conducting shell of internal and external radii a and b is cut into two by a plane at a distance p from the centre. A charged sphere is placed inside the shell and concentric with it. The two sections of the shell are put together in electrical contact but the shell as a whole is uncharged and insulated. Prove that the two sections tend to separate if

$$p > \frac{ab}{(a^2+b^2)^{1/2}}.$$

The charges on the surfaces of the spheres are Q, $-Q$, Q as shown in Fig. 3.11. These give surface densities, on $r = a$, $\sigma_a = -Q/(4\pi a^2)$; on $r = b$, $\sigma_b = Q/(4\pi b^2)$. Hence, by eqn. (2.31), the outward force on a unit area of each face is

$$\text{on } r = a, \quad \frac{Q^2}{32\pi^2 a^4 \varepsilon_0}, \quad \text{on } r = b, \quad \frac{Q^2}{32\pi^2 b^4 \varepsilon_0}.$$

Using spherical polar coordinates (r, θ, ψ) with Oz, which is perpendicular to the plane of section, as axis, the elements of area on the faces $r = a$, $r = b$ are respectively subject to forces

$$\frac{Q^2}{32\pi^2 a^2 \varepsilon_0} \sin\theta\, d\theta\, d\psi, \quad \frac{Q^2}{32\pi^2 b^2 \varepsilon_0} \sin\theta\, d\theta\, d\psi$$

out of the metal. We integrate these forces to find the resultant force on the metal above the plane of section, taking resolutes X, Y perpendicular to Oz and Z parallel to Oz:

$$X = \iint \frac{Q^2}{32\pi^2 b^2 \varepsilon_0} \sin\theta \cos\psi \sin\theta\, d\theta\, d\psi - \iint \frac{Q^2}{32\pi^2 a^2 \varepsilon_0} \sin\theta \cos\psi \sin\theta\, d\theta\, d\psi,$$

$$Y = \iint \frac{Q^2}{32\pi^2 b^2 \varepsilon_0} \sin\theta \sin\psi \sin\theta\, d\theta\, d\psi - \iint \frac{Q^2}{32\pi^2 a^2 \varepsilon_0} \sin\theta \sin\psi \sin\theta\, d\theta\, d\psi,$$

$$Z = \iint \frac{Q^2}{32\pi^2 b^2 \varepsilon_0} \cos\theta \sin\theta\, d\theta\, d\psi - \iint \frac{Q^2}{32\pi^2 a^2 \varepsilon_0} \cos\theta \sin\theta\, d\theta\, d\psi.$$

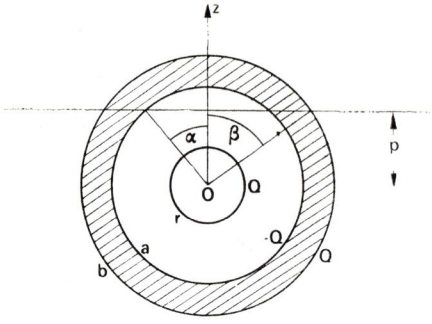

Fig. 3.11

In each of the above expressions the first integral is taken over the cap of the sphere $r = a$ and the second over the cap of $r = b$. Both X, Y vanish on integration w.r. to ψ, because of the cylindrical symmetry. Also

$$Z = 2\pi \frac{Q^2}{32\pi^2\varepsilon_0} \left\{ \frac{1}{b^2} \int_0^\beta \cos\theta \sin\theta \, d\theta - \frac{1}{a^2} \int_0^\alpha \cos\theta \sin\theta \, d\theta \right\}$$

$$= \frac{Q^2}{16\pi\varepsilon_0} \left\{ \frac{1}{2b^2}(1-\cos^2\beta) - \frac{1}{2a^2}(1-\cos^2\alpha) \right\}.$$

But $p = b\cos\beta = a\cos\alpha$. Therefore

$$Z = \frac{Q^2}{32\pi\varepsilon_0} \left\{ \left(\frac{1}{b^2} - \frac{1}{a^2}\right) - p^2\left(\frac{1}{b^4} - \frac{1}{a^4}\right) \right\}.$$

The two halves will separate if the force Z acting on the upper portion is positive, i.e. separation occurs if

$$p^2\left(\frac{1}{b^4} - \frac{1}{a^4}\right) < \frac{1}{b^2} - \frac{1}{a^2},$$

i.e.
$$p^2\left(\frac{1}{b^2} + \frac{1}{a^2}\right) > 1, \quad p > \frac{ab}{(a^2+b^2)^{1/2}}.$$

Calculation of the electrostatic energy in terms of p and differentiation w.r. to p is not a practicable method in this problem because of the difficulty in finding the mutual energy of the two parts of the charge distribution.

Displacements at Constant Potential

In deriving the form (3.17) for the ponderomotive force as the derivative of the energy, it was tacitly assumed that in the displacement specified by δx the sole expenditure of energy was by the ponderomotive forces; no energy was supplied from any other source. This assumption requires the charges in the distribution to remain unaltered in total amount; they may redistribute themselves on the conductors as a result of the displacement, and consequ-

§ 3.4 CONDUCTORS IN THE ELECTROSTATIC FIELD

ently the potentials of the conductors may alter as a result of the displacement. However, the conclusion is valid only if the charges are unaltered.

Nevertheless, by connecting each conductor to a suitable battery it would be possible to maintain its potential constant during the displacement. To do this each battery must supply additional charge to the system at the appropriate potential and so supplies energy to the system. In this case eqn. (3.16) must be modified to take account of this additional energy.

Consider a system of charged conductors; if the batteries supply charge δQ_i to the conductor at potential V_i in the course of a displacement, the net gain in energy of the system is

$$\delta W = \sum_i V_i \delta Q_i - \sum X \delta x, \qquad (3.26)$$

instead of that given in eqn. (3.16). When a displacement is made at constant charge we find $-(\partial W/\partial x)_e = X$ since $\delta Q_i = 0$. But $W = \frac{1}{2}\sum Q_i V_i$ and therefore

$$\sum V_i \delta Q_i + \sum Q_i \delta V_i = 2\delta W.$$

Hence we can also write

$$\delta W = 2\delta W - \sum_i Q_i \delta V_i - \sum X \delta x.$$

Therefore
$$\delta W = \sum Q_i \delta V_i + \sum X \delta x. \qquad (3.27)$$

If, now, we consider a displacement, specified by δx, made so that the potentials remain constant, then

$$\delta W = \sum X \delta x \quad \text{or} \quad \left(\frac{\partial W}{\partial x}\right)_V = +X. \qquad (3.28)$$

Note the difference of sign.

We can express $W = \frac{1}{2}\sum Q_i V_i$ in two forms

$$\tfrac{1}{2}\sum p_{ij} Q_i Q_j \quad \text{or} \quad \tfrac{1}{2}\sum C_{ij} V_i V_j.$$

In the first of these expressions the dependence of W on the generalized coordinates is through the coefficients p_{ij}, the dependence on the charges being explicitly shown as the quadratic form in Q_i. Therefore

$$X = -\frac{1}{2}\sum_{i,j} \frac{\partial p_{ij}}{\partial x} Q_i Q_j. \qquad (3.29)$$

In the second form W depends on the generalized coordinates through the coefficients C_{ij}, the dependence on the potentials being the quadratic form in V_i. There is no explicit dependence on charges. Therefore

$$X = +\frac{1}{2}\sum_{i,j} \frac{\partial C_{ij}}{\partial x} V_i V_j. \qquad (3.30)$$

Example. The energy stored in a charged parallel plate capacitor, in which the separation of the plates is x, which is small, is

$$W = \frac{1}{2} Q^2/C = \frac{1}{2} \frac{Q^2 x}{\varepsilon_0 \alpha}$$

$$= \frac{1}{2} CV^2 = \frac{1}{2} \frac{\varepsilon_0 \alpha V^2}{x}.$$

[See example 1, p. 85.]

From the first of these, $Q^2/(2C)$, the force tending to increase x is given by eqn. (3.29),

$$F = -\left(\frac{\partial W}{\partial x}\right)_Q = -\frac{1}{2} \frac{Q^2}{\varepsilon_0 \alpha} = -\alpha\left(\frac{\sigma^2}{2\varepsilon_0}\right). \qquad (1)$$

From the second of these expressions, $\frac{1}{2}CV^2$, the force tending to increase x is given by

$$F = +\left(\frac{\partial W}{\partial x}\right)_V = -\frac{\varepsilon_0 \alpha V^2}{2x^2}.$$

But
$$V = \frac{Q}{C} = \frac{Qx}{\varepsilon_0 \alpha}.$$

Therefore
$$F = -\frac{Q^2}{2\varepsilon_0 \alpha} = -\alpha\left(\frac{\sigma^2}{2\varepsilon_0}\right). \qquad (3)$$

These two (identical) expressions for the force show an attraction (because of the minus sign) between the plates, and correspond to a force $\sigma^2/(2\varepsilon_0)$ acting on unit area of each plate. This agrees with result (2.31).

This force of attraction between the plates of a charged parallel plate capacitor is the basis of the attracted disc electrometer. By measuring the force of attraction and x, the potential difference may be deduced from eqn. (2).

Exercises 3.4

1. A conducting sphere of radius a is surrounded by a concentric conducting spherical shell of inner and outer radii b, c respectively ($c > b > a$). Find the electrostatic energy of the system in the following two cases:
   (i) Sphere at potential V_1, shell earthed;
   (ii) Sphere earthed, shell at potential V_2.

2. A spherical conductor of radius a is maintained at constant potential V. It is surrounded by a concentric spherical conducting shell of radius b, which is insulated and carries a charge e. Show that, at a point at distance r from the centre of the spheres ($a \leqslant r \leqslant b$), the potential is

$$\frac{1}{4\pi\varepsilon_0}\left\{\frac{e}{b} + \left(V - \frac{e}{b}\right)\frac{a}{r}\right\}$$

Find the change in the electrostatic energy of the system when the radius of the inner sphere is changed from a to a'.

3. An insulated spherical conductor, formed of two hemispherical shells in contact, whose inner and outer radii are b and b', has within it a concentric spherical conductor of radius a and without it another concentric spherical conductor of which the internal radius is c. These two conductors are earth connected and the middle one receives a charge. Show that the two shells will not separate if

$$2ac > bc + b'a.$$

3.5 Criticisms of the expression for energy

There are several objections which can be made of the above energy calculations. Previously, to avoid obscuring the ideas with too much detail, some points were deliberately glossed over. We now examine certain of these points in detail.

The first group of difficulties arises because of the existence of singularities in the field. The expressions for the energy of a spherical shell carrying a charge Q and a solid sphere carrying a charge Q [examples 3 and 4, p. 96] are

$$\frac{1}{8\pi\varepsilon_0}\frac{Q^2}{a}, \quad \frac{3}{20\pi\varepsilon_0}\frac{Q^2}{a}$$

respectively. Suppose now we consider the limit $a \to 0$, Q remaining finite and not zero; this limit gives a point charge Q. Both the above expressions tend to infinity. Hence it requires infinite energy to "compress" a charge into a point charge. We have ignored these (infinite) energies entirely. Similarly we ignored any energy involved in forming a dipole; in this process, in the limit, two equal and opposite point charges have been made to coincide. We must therefore qualify the conception of "infinite dispersal"; when charges are collected from infinity and end as point charges or as dipoles we must assume that the "compression" has already been done at infinity, before the charges are moved into position. Unless we ignore the energy required for this compression, we must contemplate infinite energy whenever there is a singularity (point charge) in the field. The expression in eqn. (3.22) for the energy of a set of point charges ignores the self-energy of each point charge and gives only the mutual energy.

The second difficulty concerns the method of building up the field to obtain the expression for the energy. We use four different methods of building up the field (cf. the method for point charges and the method for conductors) but, when the charges move, we assume that the change of energy depends solely upon change of position. This change of energy occurs in a process which differs from the process of building up the field.

When we write down eqn. (3.17) giving the ponderomotive force as the derivative of a potential energy function we are *assuming* that the energy required to build up a configuration of charge is a function only of the coordinates describing this final configuration; and does not depend upon the method or sequence of operations followed in the building process. We *assume* that dW given, for example, in obtaining eqn. (3.24) is a perfect differential of the coordinates describing the final configuration (see *C.M.*, vol. 4,

§ 4.4). We cannot give much justification for this assumption except that in many simple cases the expressions obtained do agree with the earlier theory and with experiment.

The third difficulty is not so very significant in the present case but arises when we consider materials (dielectrics) other than a vacuum in which a field is established. Thus far we have assumed that we can keep distinct electrostatic energy and energy from other sources. For example, changes of temperature involve changes of energy. How do we separate this type of energy change from changes of electrostatic energy? The law of conservation of energy does *not* state that energies of "different kinds" are conserved separately; it states that energy may have different forms, electrostatic, thermal, chemical, and other forms, and that the total amount of all kinds of energy in an isolated system remains fixed. The science of thermodynamics is concerned with the laws of energy and, strictly, the questions we are now discussing should be investigated in accordance with the principles of thermodynamics. This is beyond the scope of the present volume. Since we are concerned, at present, with the properties of a vacuum we presume that complications of thermal energy cannot arise. Subsequently, when we consider polarizable media, if we assume that all changes take place isothermally (without change of temperature), then the energy changes discussed are those in what is called "available energy" in thermodynamics.

According to the definitions adopted so far we can only say that, when infinitesimal additions $\delta\varrho$, $\delta\sigma$, δQ_i, δQ_l are made to the charges of the system, the work done is

$$\delta W = \iiint \delta\varrho V \, d\tau + \iint \delta\sigma V \, dS + \sum_i \delta Q_i V_i + \sum_l \delta Q_l V_l. \tag{3.31}$$

The changes must be *infinitesimal* so that δW is correct only to the first order; changes in the field quantities arising from the introduction of new charges affect δW in the second order. Further, the potential V must be the potential due to the *remaining* charges of the field, i.e. excluding the point charge Q_l which is being augmented by δQ_l. (If V_l included the potential of the point charge Q_l which is being augmented, then the product $V_l \, \delta Q_l$ would be infinite, because it would include part of the (infinite) energy of compressing the charge $Q_l + \delta Q_l$ into a point charge.) The expression (3.31) makes none of the assumptions discussed above; but we can only "integrate" δW to give a function W by making assumptions such as those discussed in the preceding paragraph.

3.6 The location of electrostatic energy

The field outlook regards the electrostatic energy discussed above as located in the field. We seek now to obtain expressions for the way in which this energy is distributed. We consider a region of space bounded by a surface S_0 which encloses all volume and surface distributions of charge ϱ, σ; the latter is distributed on surfaces K, usually conductors, which are enveloped by

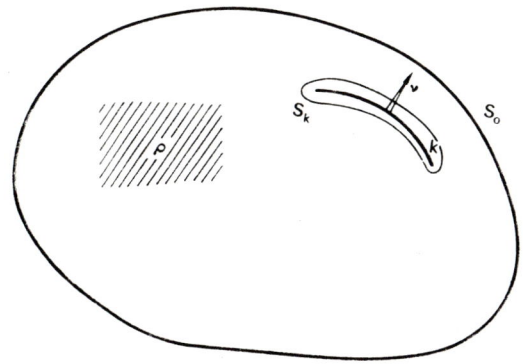

Fig. 3.12

surfaces S_K (see Fig. 3.12). We are here using the technique introduced in § 2.9. The total energy W of the whole system of charges and conductors is

$$W = \frac{1}{2} \iiint \varrho V \, d\tau + \frac{1}{2} \iint \sigma V \, dS,$$

where the surface integral is taken over the surfaces K carrying the surface charges. Since div $\boldsymbol{D} = \varrho$ for all points, we may write

$$\iiint_{S_0} \varrho V \, d\tau = \iiint_{S_0} V \, \text{div } \boldsymbol{D} \, d\tau = \iiint_{S_0} \{\text{div }(V\boldsymbol{D}) - \boldsymbol{D} \cdot \textbf{grad } V\} \, d\tau.$$

This integral is taken over the whole region inside S_0 and outside the conductors and the surfaces S_K enveloping K. Then

$$\iint_{S_K} V \boldsymbol{D} \cdot \hat{\boldsymbol{n}} \, dS \to \iint_K V \hat{\boldsymbol{v}} \cdot (\boldsymbol{D}_- - \boldsymbol{D}_+) \, dS = -\iint_K \sigma V \, dS,$$

when the enveloping surface S_K contracts to coincide in the limit with K. Hence

$$\frac{1}{2} \iiint \varrho V \, d\tau + \frac{1}{2} \iint_K \sigma V \, dS - \frac{1}{2} \iiint_{S_0} \boldsymbol{D} \cdot \boldsymbol{E} \, d\tau = \frac{1}{2} \iint_{S_0} V D_n \, dS. \quad (3.32)$$

The surface S_0 used here encloses all the charge which is the source of the field, but it does not enclose the whole of the region of the field. We presume that some of the total energy of the system is located at points outside S_0 where $\boldsymbol{E} \neq 0$. We therefore interpret the surface integral on the right-hand side of eqn. (3.31) as giving the difference between the total energy of the system and the energy located inside S_0; this is equivalent to regarding the energy as distributed in the field with volume density

$$\tfrac{1}{2}\boldsymbol{D}\cdot\boldsymbol{E} = \tfrac{1}{2}\varepsilon_0 E^2.$$

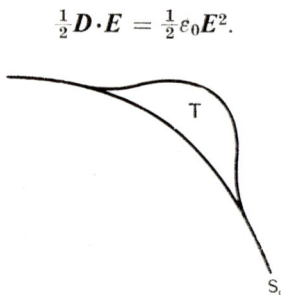

Fig. 3.13

This interpretation is confirmed if we make an arbitrary deformation of S_0 to S_0' enclosing an extra volume T which does not include any charge (Fig. 3.13). The increase in the right-hand side of eqn. (3.32) is

$$\frac{1}{2}\iint_{S_0'} VD_n \, \mathrm{d}S - \frac{1}{2}\iint_{S_0} VD_n \, \mathrm{d}S = \frac{1}{2}\iiint_T \operatorname{div}(V\boldsymbol{D}) \, \mathrm{d}\tau$$
$$= \frac{1}{2}\iiint_T \{V \operatorname{div} \boldsymbol{D} + \boldsymbol{D}\cdot\operatorname{grad} V\} \, \mathrm{d}\tau = -\frac{1}{2}\iiint_T \boldsymbol{D}\cdot\boldsymbol{E} \, \mathrm{d}\tau.$$

But this is just the extra amount of energy that would be enclosed by the deformation if the energy were distributed with the stated volume density. Since the deformation is arbitrary we can therefore regard the energy as located in the field with this density.

It is sometimes suggested that, since the addition to the energy density of any scalar field whose integral over all space vanishes will not alter the total energy, this expression $\tfrac{1}{2}\varepsilon_0 E^2$ is not unique and that the energy density is indeterminate. If we insist that the energy density, which we denote for the moment by ψ, must be an expression depending only on the vectors $\boldsymbol{E}, \boldsymbol{D}$ (at least in the case of a vacuum for which $\boldsymbol{D} = \varepsilon_0 \boldsymbol{E}$) this indeterminacy is removed. For the above discussion shows that, if we deform S_0 so as not to cross any charge, the difference between the two volume integrals must be expressible as a surface integral, and there must be a vector A (say) such

$$\iiint_T \psi \, d\tau = \iint_T A \cdot dS,$$

that whenever $\operatorname{div} D = 0$ and $\operatorname{curl} E = 0$, inside T. Since ψ is a scalar function of E and is invariant for rotation of axes it can depend only on E^2, and the above condition can be written, since T is arbitrary,

$$\psi = \operatorname{div} A,$$

and it is clear that A may depend at most on V, the potential, and E. Therefore it must have the form

$$A = \chi(V, E^2)E,$$

where χ is a scalar function of the two parameters V, E^2. We take the divergence of this relation, remembering that $\operatorname{div} E = 0$, and obtain

$$\psi(E^2) = \operatorname{div} A = -\frac{\partial \chi}{\partial V} E^2 + 2 \frac{\partial \chi}{\partial (E^2)} [(E \cdot \nabla E) \cdot E].$$

The second term on the right-hand side must involve derivatives of E, contrary to our hypothesis, unless $\partial \chi / \partial (E^2) = 0$, so that $\chi = \chi(V)$ and depends on V alone. Moreover, $\partial \chi / \partial V$ must itself be independent of V, or ψ would depend upon V as well as upon E^2. We therefore conclude that ψ must be a simple multiple of E^2, which was the form found in eqn. (3.32), the multiplier being $\frac{1}{2}\varepsilon_0$.

Example 1. A spherical shell with total charge Q.

$$\text{For } r < a, \quad \varepsilon_0 E = D = 0;$$
$$\text{for } r \geq a, \quad \varepsilon_0 E = D = \frac{Qr}{4\pi r^3}.$$

The energy density is $\dfrac{Q^2}{32\varepsilon_0 \pi^2 r^4}$. Therefore

$$W = \int_a^\infty \frac{Q^2}{32\varepsilon_0 \pi^2 r^4} \cdot 4\pi r^2 \, dr = \frac{Q^2}{8\pi\varepsilon_0} \int_a^\infty \frac{dr}{r^2} = \frac{Q^2}{8\pi\varepsilon_0} \left[-\frac{1}{r}\right]_a^\infty = \frac{Q^2}{8\pi\varepsilon_0 a},$$

as it should be by use of the formula $W = \frac{1}{2}QV$.

Now consider the energy inside a concentric sphere of radius b. (This is the surface S_0 used in the discussion.) The volume integral contribution is

$$\iiint_{S_0} \frac{1}{2}\varepsilon_0 E^2 \, d\tau = \frac{1}{2\varepsilon_0} \iiint \frac{Q^2 r^2 \, d\tau}{16\pi^2 r^6} = \frac{Q^2}{8\pi\varepsilon_0} \int_a^b \frac{dr}{r^2} = \frac{Q^2}{8\pi\varepsilon_0}\left(\frac{1}{a} - \frac{1}{b}\right).$$

The contribution from the surface integral is

$$\frac{1}{2} \iint_{S_0} VD_n \, dS = \frac{1}{2} \iint \frac{Q}{4\pi\varepsilon_0 b} \cdot \frac{Q}{4\pi b^2} b^2 \sin\theta \, d\theta \, d\psi$$

$$= \frac{Q^2}{32\pi^2 \varepsilon_0 b} \iint \sin\theta \, d\theta \, d\psi = \frac{Q^2}{8\pi\varepsilon_0 b}.$$

These two contributions together give the same total energy.

Example 2. A spherical volume of total charge Q uniformly distributed.

$$\text{For} \quad r < a, \quad \varepsilon_0 E = D = \frac{Q}{4\pi a^3} r;$$

$$\text{for} \quad r \geq a, \quad \varepsilon_0 E = D = \frac{Qr}{4\pi r^3},$$

Therefore
$$W = \frac{1}{2} \int_0^a \frac{Q^2 r^2}{16\pi^2 \varepsilon_0 a^6} 4\pi r^2 \, dr + \frac{1}{2} \int_a^\infty \frac{Q^2 r^2}{16\pi^2 \varepsilon_0 r^6} 4\pi r^2 \, dr$$

$$= \frac{Q^2}{8\pi\varepsilon_0 a^6} \int_0^a r^4 \, dr + \frac{Q^2}{8\pi\varepsilon_0 a} = \frac{3Q^2}{20\pi\varepsilon_0 a}.$$

Example 3. The electrostatic energy of the field of a number of charged conductors is

$$W = \frac{1}{2} \iiint \mathbf{E} \cdot \mathbf{D} \, d\tau$$

taken through the volume outside all conductors. But

$$\iiint \mathbf{E} \cdot \mathbf{D} \, d\tau = -\iiint \mathbf{D} \cdot \mathbf{grad}\, V \, d\tau = \iiint V \, \mathrm{div}\, \mathbf{D} \, d\tau - \iiint \mathrm{div}\,(V\mathbf{D}) \, d\tau.$$

Since there is no volume charge, $\mathrm{div}\, \mathbf{D} = 0$ at all points of the field and so

$$\iiint (\mathbf{E} \cdot \mathbf{D}) \, d\tau = \sum_i \iint_{S_i} V D_n \, dS + \iint_{S_0} V D_n \, dS,$$

where D_n is the resolute of $\mathbf{D}$ drawn *into* the field. (This is the reason for omitting the minus sign.) On each conductor $V = V_i$ ($=$ constant) and $D_n = \sigma$. Also the integral over S_0 tends to zero. Therefore

$$\iiint \mathbf{E} \cdot \mathbf{D} \, d\tau = \sum_i V_i \iint_{S_i} \sigma \, dS = \sum_i V_i Q_i.$$

Therefore
$$W = \frac{1}{2} \iiint \mathbf{E} \cdot \mathbf{D} \, d\tau = \frac{1}{2} \sum_i V_i Q_i$$

as before (p. 95).

§ 3.6 CONDUCTORS IN THE ELECTROSTATIC FIELD

Miscellaneous Exercises III

1. $S_1, \ldots, S_n$ are closed surfaces, none of which encloses another. A system of charges, none of which lies outside the surfaces, produces a potential V in the infinite external region. A different system of charges produces a potential V'. Prove that

$$\sum_{i=1}^{n} \int_{S_i} V \frac{\partial V'}{\partial n} \, dS = \sum_{i=1}^{n} \int_{S_i} V' \frac{\partial V}{\partial n} \, dS,$$

where $\partial/\partial n$ denotes differentiation along the normal drawn outward from the surface. Hence show that, if conductors $C_1, \ldots, C_n$ have potentials $V_1, \ldots, V_n$ when they carry charges $e_1, \ldots, e_n$, and have potentials $V_1', \ldots, V_n'$ when they carry charges $e_1', \ldots, e_n'$, then

$$\sum_{i=1}^{n} e_i V_i' = \sum_{i=1}^{n} e_i' V_i.$$

State and prove the modified theorem applicable to a system of point charges and conductors.
A point charge e is placed at a distance c from the centre of an uncharged conducting sphere of radius a. Find the potential of the sphere.

2. A system of n conductors consists of n closed conducting surfaces $C_1, C_2, \ldots, C_n$ of which C_n completely surrounds C_{n-1}, C_{n-1} completely surrounds C_{n-2}, and so on. Show that the coefficients of potential of the system are such that

$$p_{ij} = p_{ji} = p_{jj} \quad (i \leq j; j = 1, 2, \ldots, n),$$

and that the coefficients of capacity q_{ij} vanish when $|i-j| > 1$.

3. Show by using Green's reciprocal theorem that the two expressions in example 5 of p. 97 for the mutual energy of two (rigid) distributions of charge are equal.

4. Prove (i) that if a conductor, insulated in free space and raised to unit potential, produces at any external point P a potential denoted by (P), then a unit charge placed at P in the presence of this conductor uninsulated will induce on it a charge $-(P)$;
(ii) that if the potential at a point Q due to the induced charge be denoted by (PQ), then (PQ) is a symmetrical function of the positions of P and Q.

5. The outer surface of a condenser is an earthed sphere of fixed radius b, and the inner surface is a concentric sphere which is maintained at constant potential V by contact with a battery. If the radius of the inner sphere increases slowly from a to a', contact with the battery being maintained, find Q and Q', the original and final charges on this sphere. Prove that

$$V(Q' - Q) = \frac{1}{2} Q'V - \frac{1}{2} QV + \int_a^{a'} (2\pi/\varepsilon_0) \sigma^2 \cdot c^2 \, dc,$$

where $4\pi c^2 \sigma$ is the charge on the inner sphere when its radius is c, and interpret this relation as an equation of energy.

6. Show that the increase ΔE_1 in the electrical energy of a system of charged conductors when they are given certain small displacements at constant potential is equal and opposite to the increase ΔE_2 in energy when the conductors are given the same displacements at constant charge. Explain why the force tending to increase a coordinate ξ is measured by $-(\partial E_2/\partial \xi)$ and not by $-(\partial E_1/\partial \xi)$.
The inner sphere (radius a) of a spherical condenser is originally at potential v and the outer sphere is kept earthed. If the inner sphere is kept at potential v, show that the system will deliver mechanical work of amount

$$\frac{\varepsilon_0 v^2 a^2 (b-c)}{8\pi (b-a)(c-a)},$$

when the outer sphere contracts from radius b to radius c. What would be the corresponding work if the inner sphere were insulated throughout the process?

7. Assuming that the electric charge Ze of an atomic nucleus is uniformly distributed inside a sphere of radius R, prove that the potential at a distance r from the centre is

$$\phi = \frac{Ze}{8\pi\varepsilon_0 R}\left\{3-\left(\frac{r}{R}\right)^2\right\} \qquad (r \leqslant R).$$

Obtain an expression for the electrostatic energy $\frac{1}{2}\int \varrho V \, d\tau$ of the nucleus and verify that it is equal to the field energy $\frac{\varepsilon_0}{2}\int E^2 \, d\tau$.

8. Two wires in the form of circles of radii a and b lie in parallel planes perpendicular to the line joining their centres, which are at a distance c apart. They carry like charges of amounts Q_1 and Q_2 respectively. Show that the mutual electrostatic energy of the system is

$$\frac{kQ_1Q_2}{2\pi\varepsilon_0\sqrt{(ab)}}\int_0^{\pi/2}\frac{d\theta}{\sqrt{(1-k^2\sin^2\theta)}},$$

where
$$k^2 = 4ab[c^2+(a+b)^2]^{-1}.$$

Find also in the form of an integral the force of repulsion between the wires.

9. A charged conductor is placed in a uniform electric field of intensity $\mathbf{E}$. Show that, at great distances from the conductor, the potential of the charges on the conductor is

$$4\pi\varepsilon_0 V = \frac{Q}{R} + \frac{\mathbf{M}\cdot\mathbf{R}}{R^3} + O\left(\frac{1}{R^3}\right),$$

where Q is the total charge on the conductor, and $\mathbf{M}$ the moment of the charges on the conductor about the origin of the vector $\mathbf{R}$.

Show that the resultant of the forces on the conductor is a force $Q\mathbf{E}$ at the origin together with a couple of moment $\mathbf{M}\times\mathbf{E}$.

10. Find the force between two parallel plates each of area A at a distance d apart in vacuo when a potential difference V is maintained between them. (Neglect edge effects.)

One plate is held rigidly horizontal and the other is suspended at its centre from a spring in which the tension is kx when it is stretched by an amount x. When there is no potential difference the plates are at a distance d apart. Show that, when a potential difference V is applied such that

$$V^2 < \frac{32\pi kd^3}{27A},$$

one of the positions of equilibrium is such that the plates are at a distance greater than $\frac{2}{3}d$.

11. An electric field *in vacuo* is due to a continuous distribution of volume charges. The charge density does not vanish anywhere within a simple closed surface S. The potential of this field is called U. A different distribution of volume charge satisfying the same conditions produces a field of potntial V. The mechanical force per unit volume exerted by the first field on the charges producing it is everywhere within S equal to the corresponding force produced by the second field. Prove that, within S, V is a function of U, and explain what this statement means.

If, then, $V = f(U)$, show that, within S,

$$\frac{df}{dU}\frac{d^2f}{dU^2}(\mathbf{grad}\ U)^2 + \left[\left(\frac{df}{dU}\right)^2 - 1\right]\nabla^2 U = 0.$$

CHAPTER 4

DIELECTRICS

4.1 The effects of a dielectric

In Chapter 1 we defined insulators as substances on which, or inside which, electric charge is unable to move continuously. Experiments show that when a piece of insulating material is introduced into a region where an electric field is already established the field is modified, the distributions of charge and the forces acting on the conductors alter, and the potentials of the conductors change. In addition, the insulator itself experiences (ponderomotive) forces; this is illustrated experimentally when dust particles and small pieces of paper are attracted to a charged body such as a gramophone record. Faraday investigated another electrostatic property of an insulating material, viz. its effect on the capacitance of a condenser. He found that when the region of the field inside a condenser (e.g. between the plates of a parallel plate condenser, or between the spheres of a spherical condenser) was filled with an insulating substance the capacitance of the condenser was multiplied by a factor K (>1) which he was able to measure. The factor K does not depend upon the charge or potential of the capacitor, but varies from one material to another. Faraday called this factor the *specific inductive capacity* of the material; the modern name is *dielectric constant*. More recently it has been found that, under a wider range of conditions than those of Faraday's investigations, the dielectric "constant" of a given material alters. However, for the steady conditions of an electrostatic field of moderate intensity, K may be taken to be independent of the field quantities.

When an insulator is situated in a field which presumably penetrates into the material of the insulator we call the substance of the insulator a dielectric. We now modify our theory to take account of the presence of dielectrics; our choice of modification is suggested by the results described above and by our rough ideas of the structure of matter as outlined in Chapter 1.

When a dielectric is situated in a field the positive and negative charges of which the dielectric is constructed experience forces in opposite directions along the lines of force of the field. The characteristic of an insulator (in contrast to a conductor) is that in response to such forces these charges are displaced, or strained, from their normal equilibrium positions and take up new equilibrium positions. (In a conductor the charges move until any excess charge is situated on the surface of the conductor.) In these new positions of equilibrium a pair of positive and negative charges which originally neutralized each other are separated by a short distance, and so constitute a close approximation to a dipole. Thus, in addition to any simple charge which may be situated on or inside the dielectric, every element of volume becomes polarized.

Before developing the analysis in the general case we consider the effect in the special uniform case of a parallel plate condenser whose plates are separated by a distance t. We consider two situations:

(1) The plates carry charges of density $\pm \sigma$ in the absence of a dielectric. Then
$$D = \sigma = \varepsilon_0 E, \quad V = Et = \sigma t/\varepsilon_0, \quad C = \sigma/V = \varepsilon_0/t,$$
where V is the potential difference between the plates and C is the capacity of unit area of the (empty) condenser. (All vectors are parallel to the electric intensity E shown in Fig. 4.1.)

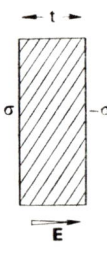

FIG. 4.1

(2) The plates carry the same charges $\pm \sigma$ as above but a slab of uniform dielectric of thickness t is placed between them. Faraday's results indicate that the capacitance becomes KC.

We assume that the polarization caused in the medium is uniform and can be represented by a vector P directed parallel to the other vectors D', E'. The Poisson charge equivalent to this polarization is a surface density $-|P|$ on the left-hand face and $+|P|$ on the right-hand face. [See § 2.12.] We apply the results of § 2.12 and write down for the field in the polarized region
$$D' = \sigma, \quad \varepsilon_0 E' = \sigma - P, \quad V' = E't, \quad C' = K\varepsilon_0/t.$$

§ 4.1 DIELECTRICS

Provided that the field is not too large, we make the reasonable assumption that the displacements caused in the elementary charges are proportional to the forces acting on these charges, and take place in the line of these forces, i.e. we assume

$$P = kE', \qquad (4.1)$$

where k is a constant of proportionality. Then

$$\varepsilon_0 E' = \sigma - kE', \quad (\varepsilon_0 + k)E' = \sigma = D',$$

$$V' = \frac{\sigma t}{\varepsilon_0 + k}, \quad C' = \frac{\varepsilon_0 + k}{t} = \frac{K\varepsilon_0}{t}.$$

It follows that

$$K = 1 + k/\varepsilon_0, \quad D' = \varepsilon_0 K E' = \varepsilon E', \qquad (4.2, 4.3)$$

where

$$\varepsilon = \varepsilon_0 K.$$

With the special simplifying assumptions concerning the direction of the polarization and its magnitude as given by eqn. (4.1) above, the presence of polarization in the dielectric explains Faraday's observation in this case.

The increased capacitance arising from the presence of a dielectric is accompanied by a reduced force of attraction between the plates—this force can be measured in suitable circumstances. Consequently, it is also supposed that the magnitude of the force between two point charges Q_1 and Q_2 in a medium of dielectric constant K is reduced below the value $Q_1 Q_2/(4\pi\varepsilon_0 r^2)$, the value when the charges are in a vacuum. (It is impracticable to verify this supposition experimentally when the medium is a solid.) However, we give a qualitative explanation here and discuss the matter further after the general theory has been formulated [see examples 2, 3, pp. 134–5].

If cavities are made in the medium to enclose charges Q_1 and Q_2 and then the dimensions of these cavities are reduced to zero, the Poisson charge on the wall of each cavity neutralizes a portion of the point charge so that the interaction is effectively between two charges smaller in magnitude than the true charges Q_1 and Q_2. The terms "true" charge and "effective" charge are used in this connection. The "true" charge is that placed originally by the agent building the field; in the parallel plate condenser considered above $\pm \sigma$ are the "true" charges. The "effective" charge is the remaining part of the true charge after part has been neutralized by the polarization charge; it is this remaining charge which is "effective" in interactions. In the parallel plate condenser $\pm(\sigma - |\boldsymbol{P}|)$ are the "effective" charges. We can summarize briefly (but not quite accurately) the theory, developed below, by saying that "$\boldsymbol{D}$ arises from 'true' charge, and $\varepsilon_0 \boldsymbol{E}$ arises from 'effective' charge".

4.2 The general theory of dielectrics

Now we consider any distribution of "true" charge ϱ, σ, Q_l which gives rise to a field in a region which may contain dielectric polarizable media. In contrast to the remarks made in § 3.2, here we *can* consider that both ϱ, σ may be present in the region intervening between the conductors since there is now insulating material present on which such charges may reside. We assume that at every point of the field there is a polarization P which arises in the dielectric through the action of the electric field E, but which also adds its contribution to that of the true charges in producing the resultant field at any point.

We use the formulae established in § 2.12 to give the contributions of P to the field vectors. We conclude that the field vectors satisfy

$$\text{curl } E = 0, \tag{4.4}$$

$$D = \varepsilon_0 E + P, \quad \text{div } D = \varrho. \tag{4.5, 4.6}$$

The corresponding surface relations are

$$\hat{\nu} \times (E_+ - E_-) = 0, \quad \hat{\nu} \cdot (D_+ - D_-) = \sigma. \tag{4.7, 4.8}$$

Equations (4.6) and (4.8) indicate that Gauss's theorem holds unchanged and consequently may be used in a similar manner to that in example 4, p. 43, where appeal is made to symmetry.

Since the polarization P depends upon the field, relations (4.4)–(4.6) are insufficient to determine the field. We solved the problem of the uniform field of the parallel plate condenser by *assuming* a relation $P = kE$ or the equivalent relation $D = (\varepsilon_0 + k)E = \varepsilon E$. There must be some relation additional to eqns. (4.4)–(4.6) in order that E, D, P can be determined. Such a relation between any two of these field vectors must express the response of the medium to the establishment of a field and to find this relation we consider a general relation between D and E in the form

$$D_i = a + b_1 E_1 + b_2 E_2 + b_3 E_3 + \tfrac{1}{2}(C_{11} E_1^2 + C_{22} E_2^2 + \ldots) + \ldots \tag{4.9}$$

giving a typical resolute D_i of the vector D as a Taylor expansion in terms of the resolutes of E. Here a, b_1, b_2, b_3, etc., are functions of the position of the field point to which D_i and E_i refer and, since both D and E are vectors, must transform in a suitable manner on rotation of the coordinate axes.

If $E = 0$, all the field vectors are zero; we thus deduce $a = 0$. Now we consider the case in which only the terms of first degree in E_i are important in eqn. (4.9), thereby restricting our considerations either to special media or to weak fields. With this restriction eqn. (4.9) becomes

$$D_i = \sum_{j=1}^{3} \varepsilon_{ij} E_j \quad (i = 1, 2, 3), \tag{4.10}$$

where we have replaced b_1, b_2, b_3 by ε_{i1}, ε_{i2}, ε_{i3}. Since D_i and E_j are resolutes of vectors, eqn. (4.10) implies that ε_{ij} is a second order tensor. Media for which eqn. (4.10) holds are said to be *linear*, the tensor ε_{ij} being a function of position (a tensor field; cf. a vector field). In general eqn. (4.10) implies that $\boldsymbol{D}$ and $\boldsymbol{E}$ have different directions. However, the assumption made for the parallel plate condenser was that $\boldsymbol{D}$ and $\boldsymbol{E}$ were parallel; if this is assumed to be true for all arbitrary directions of $\boldsymbol{E}$, then eqn. (4.10) becomes

$$\boldsymbol{D} = \varepsilon \boldsymbol{E}, \tag{4.11}$$

where ε is a scalar function of position. For a linear medium corresponding to eqn. (4.10) the electrical properties depend upon the direction of $\boldsymbol{E}$, or $\boldsymbol{D}$; for a medium corresponding to eqn. (4.11) these properties do not depend on the direction of $\boldsymbol{E}$. The two types of media are therefore said to be

anisotropic linear media, corresponding to eqn. (4.10)
isotropic linear media, corresponding to eqn. (4.11).

The scalar ε is called the *permittivity* of the medium and eqn. (4.11) is the generalization of eqn. (2.10) for a vacuum. A medium for which ε has the same value at every point, i.e. ε is a constant scalar, is a *uniform, isotropic linear* medium. In general we consider in this volume only linear isotropic media. In practice it is found that this restriction does not greatly limit the applicability of the theory since the actual behaviour of many dielectrics conforms very closely to that of a theoretical isotropic linear medium. If anisotropic media are included, the theory also represents the behaviour of media with crystalline structures. Nevertheless, there are systems found in practice for which the linear assumption is inadequate, but we do not consider such systems here, i.e. we do not consider the effect of the terms $C_{11}E_1^2 + C_{22}E_2^2 + \ldots$ in eqn. (4.9). (The relevant problems are very difficult!)

If we now include the relation (4.11), where ε is a scalar function of position, we find, using eqn. (4.5), that

$$\varepsilon \boldsymbol{E} = \boldsymbol{D} = \varepsilon_0 \boldsymbol{E} + \boldsymbol{P}, \quad \boldsymbol{P} = (\varepsilon - \varepsilon_0) \boldsymbol{E}. \tag{4.12}$$

Comparison with eqn. (4.1) gives $k = \varepsilon - \varepsilon_0$ and so

$$K = 1 + k/\varepsilon_0 = \varepsilon/\varepsilon_0.$$

This ratio $\varepsilon/\varepsilon_0$ is called the *relative permittivity* and is, in fact, identical with the dielectric constant. (This we show more explicitly later.) Since, in general, $K \geqslant 1$, we have $\varepsilon \geqslant \varepsilon_0 > 0$. The quantity k is sometimes called the *dielectric susceptibility*. The units of ε are the same as those of ε_0, viz. farads per metre, but K, being a ratio, is dimensionless, i.e. its value does not alter even if the basic units are changed in magnitude.

Since **curl** $E = 0$ we have $E = -\,\text{grad}\,V$ where now, from eqn. (2.15),

$$V = \iiint \frac{\varrho - \text{div}\,\boldsymbol{P}}{4\pi\varepsilon_0 r}\,d\tau + \iint \frac{\sigma - (\boldsymbol{P}_+ - \boldsymbol{P}_-)\cdot\hat{\boldsymbol{v}}}{4\pi\varepsilon_0 r}\,dS + \sum \frac{(Q_l - Q_l')}{4\pi\varepsilon_0 r}. \tag{4.13}$$

(Here Q_l' is the polarization charge which tends to neutralize Q_l.) However, these integrals do not give V unless $\boldsymbol{P}$ is known, and $\boldsymbol{P}$ is not known unless E is known [eqn. (4.12)] and E depends upon V. This difficulty resembles that encountered with fields in the presence of conductors, and is overcome in much the same manner by finding a partial differential equation for V, together with appropriate boundary conditions.

From eqn. (4.11), $\boldsymbol{D} = -\varepsilon\,\text{grad}\,V$, and eqn. (4.6) becomes

$$\text{div}\,(\varepsilon\,\text{grad}\,V) = -\varrho. \tag{4.14}$$

Also the boundary condition (4.8) gives the surface form of this relation as

$$\left(\varepsilon\frac{\partial V}{\partial v}\right)_+ - \left(\varepsilon\frac{\partial V}{\partial v}\right)_- = -\sigma. \tag{4.15}$$

As before, in the absence of double layers, we conclude from eqn. (4.14) that

$$V_+ - V_- = 0 \tag{4.16}$$

a relation which implies eqn. (4.7).

Example 1. We deduce eqn. (4.14) from the expression (4.13) which gives V as follows.
At field points which do not coincide with surface discontinuities or point charges, V given by eqn. (4.13) must satisfy

$$\nabla^2 V = -(\varrho - \text{div}\,\boldsymbol{P})/\varepsilon_0.$$

But, from eqn. (4.12),

$$\boldsymbol{P} = (\varepsilon - \varepsilon_0)\boldsymbol{E} = -(\varepsilon - \varepsilon_0)\,\text{grad}\,V.$$

Therefore, $\qquad\text{div}\,\boldsymbol{P} = -\text{div}\,(\varepsilon\,\text{grad}\,V) + \varepsilon_0\nabla^2 V.$

Therefore, $\qquad \varepsilon_0\nabla^2 V = -\varrho + \text{div}\,\boldsymbol{P} = -\varrho - \text{div}\,(\varepsilon\,\text{grad}\,V) + \varepsilon_0\nabla^2 V.$

Therefore, $\qquad\qquad \text{div}\,(\varepsilon\,\text{grad}\,V) = -\varrho.$

§4.2 DIELECTRICS

Example 2. The surface relation (4.15) can be deduced similarly.

From eqn. (4.13) across any surface of discontinuity, i.e. a surface carrying surface charge σ, and/or separating regions of different polarization P_+ and P_-, we deduce that

$$\left(\frac{\partial V}{\partial \nu}\right)_+ - \left(\frac{\partial V}{\partial \nu}\right)_- = -\frac{\sigma - (P_+ - P_-) \cdot \hat{\nu}}{\varepsilon_0}.$$

But
$$(P_+ - P_-) \cdot \hat{\nu} = \hat{\nu} \cdot [(\varepsilon - \varepsilon_0)E]_+ - \hat{\nu} \cdot [(\varepsilon - \varepsilon_0)E]_-$$
$$= \left[-(\varepsilon - \varepsilon_0)\frac{\partial V}{\partial \nu}\right]_+ - \left[-(\varepsilon - \varepsilon_0)\frac{\partial V}{\partial \nu}\right]_-.$$

Therefore,
$$\varepsilon_0 \left(\frac{\partial V}{\partial \nu}\right)_+ - \varepsilon_0 \left(\frac{\partial V}{\partial \nu}\right)_- = -\sigma - \left[(\varepsilon - \varepsilon_0)\frac{\partial V}{\partial \nu}\right]_+ + \left[(\varepsilon - \varepsilon_0)\frac{\partial V}{\partial \nu}\right]_-.$$

Therefore,
$$\left(\varepsilon \frac{\partial V}{\partial \nu}\right)_+ - \left(\varepsilon \frac{\partial V}{\partial \nu}\right)_- = -\sigma.$$

Example 3. A condenser is formed of two concentric conducting spheres of radii a and b ($a < b$), and the space between is filled with a substance whose dielectric constant at a point distant r from the centre is $(c+r)/r$. The outer sphere is earthed and the inner charged. Show that the capacity C of the system is given by

$$\frac{1}{C} = \frac{1}{4\pi\varepsilon_0 c} \ln\left\{\frac{b(c+a)}{a(c+b)}\right\}.$$

This is a system with spherical symmetry to which we apply Gauss's theorem using a concentric sphere of radius r as the surface G. Appealing to the symmetry, as in example 4, p. 43, we write

$$4\pi r^2 D_r = Q, \qquad a \leqslant r \leqslant b,$$

where Q is the charge on the inner sphere and D_r is the radial component of D, the other spherical components being zero. Hence at a distance r from the centre

$$\varepsilon E_r = \frac{Q}{4\pi r^2}, \quad E_\theta = E_\psi = 0, \quad \text{i.e.} \quad E_r = \frac{Q}{4\pi\varepsilon_0 r(c+r)} = -\frac{dV(r)}{dr}.$$

Hence
$$\frac{dV}{dr} = \frac{Q}{4\pi\varepsilon_0 c}\left\{\frac{1}{c+r} - \frac{1}{r}\right\},$$

$$V = \frac{Q}{4\pi\varepsilon_0 c} \ln\left(\frac{c+r}{r}\right) + A.$$

But $V = V_1$ where $r = a$, $V = 0$ where $r = b$. Therefore

$$V_1 = \frac{Q}{4\pi\varepsilon_0 c} \ln\left\{\frac{b(a+c)}{a(b+c)}\right\}.$$

Hence the capacitance C is given by

$$\frac{V_1}{Q} = \frac{1}{C} = \frac{1}{4\pi\varepsilon_0 c} \ln\left\{\frac{b(a+c)}{a(b+c)}\right\}.$$

Example 4. A condenser consists of two long coaxial conducting cylinders with a coaxial cylindrical shell of dielectric, of specific inductive capacity K, placed between them. The radius of the inner cylinder is a; the inner and outer radii of the dielectric shell are b and c;

and the inner radius of the outer shell is d. Neglecting end effects, prove that the capacity per unit length of the condenser is

$$\frac{2\pi\varepsilon_0}{\left\{\ln\left(\frac{b}{a}\right)+\frac{1}{K}\ln\left(\frac{c}{b}\right)+\ln\left(\frac{d}{c}\right)\right\}}.$$

When such a pair of coaxial cylinders is used as a condenser the outer one is earthed and the inner charged (unless otherwise indicated).

The phrase "neglecting end effects" implies that we may regard the cylinders as of infinite length and take the field as a two-dimensional field. The arrangement has cylindrical symmetry and is uniform in the direction of the axis so we apply Gauss's theorem using a concentric cylinder of unit axial length and radius r.

By an appeal to the cylindrical symmetry we deduce that

$$2\pi r D_r = e, \qquad a \leqslant r \leqslant d. \tag{1}$$

Here e is the charge on unit length of the inner cylinder. We emphasize that, despite the non-uniformity of the dielectric, eqn. (1) applies for *all* values of r inside or outside the dielectric.

We therefore deduce, for the following ranges,

$$a \leqslant r \leqslant b \qquad b \leqslant r \leqslant c \qquad c \leqslant r \leqslant d$$

$$E_r = \frac{e}{2\pi\varepsilon_0 r}, \qquad \frac{e}{2\pi\varepsilon_0 K r}, \qquad \frac{e}{2\pi\varepsilon_0 r}.$$

We integrate the relation $E_r = -dV/dr$ to obtain

$$V = -\frac{e}{2\pi\varepsilon_0}\ln r + A_1, \qquad -\frac{e}{2\pi\varepsilon_0 K}\ln r + A_2, \qquad -\frac{e}{2\pi\varepsilon_0}\ln r + A_3.$$

To determine the constants A_1, A_2, A_3 we use the boundary conditions:

Where

$r = a, \quad V = V_1;$ $\qquad\qquad V_1 = -\dfrac{e}{2\pi\varepsilon_0}\ln a + A_1;$

$r = b, \quad V \text{ is continuous}; \qquad -\dfrac{e}{2\pi\varepsilon_0}\ln b + A_1 = -\dfrac{e}{2\pi\varepsilon_0 K}\ln b + A_2;$

$r = c, \quad V \text{ is continuous}; \qquad -\dfrac{e}{2\pi\varepsilon_0 K}\ln c + A_2 = -\dfrac{e}{2\pi\varepsilon_0}\ln c + A_3;$

$r = d, \quad V = 0; \qquad\qquad -\dfrac{e}{2\pi\varepsilon_0}\ln d + A_3 = 0.$

Therefore by addition

$$V_1 - \frac{e}{2\pi\varepsilon_0}\left(\ln b + \frac{1}{K}\ln c + \ln d\right) = -\frac{e}{2\pi\varepsilon_0}\left(\ln a + \frac{1}{K}\ln b + \ln c\right),$$

i.e. $\qquad V_1 = \dfrac{e}{2\pi\varepsilon_0}\left\{\ln\left(\dfrac{b}{a}\right)+\dfrac{1}{K}\ln\left(\dfrac{c}{b}\right)+\ln\left(\dfrac{d}{c}\right)\right\}.$

Hence the capacitance (of unit length) is given by C where

$$\frac{V_1}{e} = \frac{1}{C} = \frac{1}{2\pi\varepsilon_0}\left\{\ln\left(\frac{b}{a}\right)+\frac{1}{K}\ln\left(\frac{c}{b}\right)+\ln\left(\frac{d}{c}\right)\right\}.$$

§ 4.2 DIELECTRICS

Example 5. A condenser is formed of two parallel plates, distant h apart, one of which is at zero potential. The space between the plates is filled with a dielectric whose dielectric constant increases uniformly from one plate to the other. Show that the capacity per unit area is

$$\varepsilon_0(K_2-K_1)/\{h \ln (K_2/K_1)\},$$

where K_1 and K_2 are the values of the dielectric constants at the surfaces of the plates. The inequalities of distribution at the edges of the plates are neglected.

Neglect of "inequalities of distribution at the edges" implies that we regard the field as uniform everywhere between the plates. Since the dielectric constant increases uniformly from one plate to the other we can write

$$K = K_1+(K_2-K_1)(z/h),$$

where K is the dielectric constant at a distance z from the first plate.

We apply Gauss's theorem using a cuboid surface G which intercepts unit area on the first plate, has its side faces parallel to the field, and is closed by planes in the material of the first plate and at distance z from the surface of this plate [cf. example 2, p. 42]. Then

$$D_z = \sigma, \quad 0 \leqslant z \leqslant h.$$

Therefore

$$E_z = -\frac{dV}{dz} = \frac{\sigma}{\varepsilon_0 K_1+\varepsilon_0(K_2-K_1)(z/h)},$$

$$V = -\frac{\sigma h}{\varepsilon_0(K_2-K_1)} \ln \{K_1 h+(K_2-K_1)z\}+A.$$

Where $z = 0$, $V = V_1$; where $z = h$, $V = 0$. Therefore

$$V_1 = \frac{\sigma h}{\varepsilon_0(K_2-K_1)} \ln \left(\frac{K_2}{K_1}\right).$$

Hence the capacitance of unit area is

$$C = \varepsilon_0(K_2-K_1)/\{h \ln (K_2/K_1)\}.$$

Note. In each of the examples 3, 4 and 5 above we deduced an expression for D which was valid in all regions despite the variations in dielectric constant. We were able to do this only because the symmetry of the system required the direction of D to be perpendicular to the surfaces of discontinuity. Hence the boundary conditions stating the continuity of the *normal component* of D in these cases implied the continuity of D itself. In systems with less symmetry we cannot usually deduce that D itself is continuous. [The tangential boundary conditions relate the components of E.]

Exercises 4.2

1. Find the capacity of a condenser consisting of two concentric conducting spherical shells of radii a, $2a$ respectively, the space between the shells being free from dielectric. Show that it is possible to double the capacity of the condenser by inserting a slab of uniform dielectric having spherical boundaries concentric with the conducting spheres, provided that the dielectric constant K is not less than 2.

2. A capacitor consists of two parallel equal conducting plates, each of area S, arranged opposite each other at a distance d apart. A parallel slab of dielectric of thickness d' and volume W, with $d' \leqslant d$ and $W \leqslant Sd'$, is placed between the plates at a distance $a \leqslant d-d'$ from one of the plates. The slab has dielectric constant K. Neglecting end effects,

find the capacity of the system, verifying that it is independent of a. Show that if $K > 1$, and W is fixed, for varying d' the capacity is greatest when $d' = d$, being then equal to

$$\frac{\varepsilon_0 S}{d}\left[1+(K-1)\frac{W}{Sd}\right].$$

3. A spherical condenser consists of two concentric conducting spheres of radii a, b $(b > a)$. A spherical shell of dielectric K extends from the inner sphere to a distance c $(<b)$ from the centre. The inner sphere is insulated and receives a charge q and the outer is earthed. Prove that the potential inside the dielectric is

$$\frac{1}{4\pi\varepsilon_0}\left(\frac{q}{Kr}-\frac{q}{Kc}+\frac{q}{c}-\frac{q}{b}\right),$$

where r is the distance of a point from the centre.
Find also the capacity of the condenser.

4. If a spherical conductor, of radius a with no other conductor in the neighbourhood, is coated with a uniform thickness d of dielectric whose specific inductive capacity varies inversely as the square of the distance from the centre, show that the capacity of the conductor is increased in the ratio

$$Ka(a+d): Ka^2+ad+d^2,$$

where K is the value of the s.i.c. at the surface.

5. A condenser is formed of two parallel plates, distant h apart, one of which is at zero potential. The space between the plates is filled with a dielectric whose inductive capacity k varies continuously from one plate to the other. Show that the capacity per unit area is

$$\varepsilon_0\left(\int_0^h \frac{dx}{k}\right)^{-1},$$

where x is a coordinate measured in a direction normal to the plates. The inequalities of distribution at the edges of the plates are neglected.

6. An electrostatic system consists of two concentric conducting circular cylinders of radii a, b $(a < b)$. The region between the cylinders is divided into $2n$ concentric layers of equal thickness, and alternate layers are filled with homogeneous materials of dielectric constants K_1, K_2, respectively. Determine the capacity per unit length of the system, and show that, as $n \to \infty$, the layers are equivalent to a single homogeneous material of dielectric constant $2K_1 K_2/(K_1+K_2)$.

4.3 The uniqueness and reciprocal theorems

The general electrostatic field contains polarizable media, in addition to conductors and charge distributions ϱ, σ, Q_l. Consequently we must modify the investigation of § 3.2. Instead of the conditions (1)–(5) of § 3.2, pp. 74–5, the potential function V, when polarizable media are present, has to satisfy:

(1) At all points of the field, except those coinciding with surface or point charges

$$\text{div}\,(\varepsilon\,\mathbf{grad}\,V) = -\varrho,$$

where the scalar $\varepsilon \geqslant \varepsilon_0 > 0$.

§ 4.3 DIELECTRICS

(2) V satisfies the discontinuity relation

$$\left(\frac{\partial V}{\partial \nu}\right)_+ - \left(\frac{\partial V}{\partial \nu}\right)_- = -\frac{\sigma}{\varepsilon}$$

at all points of a surface carrying a rigid surface distribution of charge σ. In the absence of double layers the potential function is continuous across layers of surface charge.

(3) V has a singularity at every point charge Q_l such that

$$V = \frac{Q_l}{4\pi\varepsilon s} + O(1), \quad |\text{grad } V| = \frac{Q_l}{4\pi\varepsilon s^2} + O(1), \quad s \to 0,$$

where s is the distance of the field point from the singularity Q_l.

(4) On each (finite) conducting boundary S_i of the field,

$$V = V_i = \text{constant}, \quad \text{and} \quad \oint \varepsilon \frac{\partial V}{\partial \nu} dS = -Q_i.$$

(5) If the region of the field extends to infinity,

$$V = O\left(\frac{1}{R}\right), \quad |\text{grad } V| = O\left(\frac{1}{R^2}\right) \quad \text{as} \quad R \to \infty.$$

These conditions ensure that a solution for V is unique. The proof relies upon Kelvin's generalization of Green's theorem; this theorem states that, if U is continuous within and on a closed surface and the scalar function of position ε is everywhere positive,

$$\oint \varepsilon U \frac{\partial U}{\partial n} dS = \iiint U \text{ div } (\varepsilon \text{ grad } U) d\tau + \iiint \varepsilon (\text{grad } U)^2 d\tau \quad (4.17)$$

(see C.M., vol. 4, p. 72, question 6). By writing $U = V' - V$, where V' and V are two functions, assumed different, which satisfy conditions (1)–(5) with given $\varepsilon, \varrho, \sigma, Q_l$, we deduce that $U = 0$, i.e. $V' = V$. (The proof follows almost exactly the lines of § 3.2.)

We can also use a similar generalization of Green's theorem to prove that Green's reciprocal theorem (3.12) holds also in the presence of (linear) polarizable media. The most general form of this theorem is as follows.

If the potential function V gives the field corresponding to a distribution of volume, surface and point charges ϱ, σ, Q_l, and charges Q_i situated on conductors at potentials V_i, and if V' gives the field corresponding to different distributions ϱ', σ', Q'_l and different charges Q'_i on the same set of conductors which are now at potentials V'_i, and the positions of conductors and polarizable media are identical in both cases, then

$$\iiint \varrho V' \, d\tau + \iint \sigma V' \, dS + \sum Q_i V'_i + \sum Q_I V'_I$$
$$= \iiint \varrho' V \, d\tau + \iint \sigma' V \, dS + \sum Q'_i V_i + \sum Q'_I V_I. \quad (4.18)$$

In this expression the volume integrals are taken throughout the whole field and the surface integrals over all surface distributions not on conductors; the potentials V_I give the values taken by V at the points occupied, in the second distribution by Q'_I, and V'_I give the values taken by V' at the points occupied by Q_I in the first arrangement. (The positions of Q_I and Q'_I need not coincide.)

Example 1. We consider a system of N conductors whose potentials and charges are related by eqns. (3.6) when they are situated in a vacuum. Now suppose that the conductors, in the same relative positions, are in a uniform uncharged medium (infinite if necessary) of permittivity ε. We find the changes which occur in the coefficients of capacitance, influence, and potential.

We assume that the potential function *in vacuo* is V where

$$V = V_i \text{ on } S_i, \quad Q_i = -\varepsilon_0 \iint_{S_i} \frac{\partial V}{\partial \nu} \, dS, \quad i = 1, \ldots, N.$$

We suppose that the potential function U corresponds to the same potentials V_i on each of the conductors, but corresponds to charges Q'_i on the respective conductors, i.e.

$$U = V_i \text{ on } S_i, \quad Q'_i = -\varepsilon \iint_{S_i} \frac{\partial U}{\partial \nu} \, dS.$$

Since the dielectric is uniform, $\nabla^2 V = 0 = \nabla^2 U$ everywhere between the conductors; both U, V satisfy the standard boundary conditions at infinity; and, on *every* finite boundary S_i, U and V take up the same values V_i; these conditions and the uniqueness theorem imply $V = U$. Moreover, we can choose the values V_i arbitrarily. Now

$$Q_i = \sum_{j=1}^{N} c_{ij} V_j \quad \text{and} \quad Q'_i = \sum_{j=1}^{N} c'_{ij} V_j$$

and, since $Q' = (\varepsilon/\varepsilon_0) Q_i$ for every conductor and V_j are arbitrary, we conclude that

$$c'_{ij} = (\varepsilon/\varepsilon_0) c_{ij} = K c_{ij}, \quad i, j = 1, 2, \ldots, N.$$

This is Faraday's result (extended slightly) concerning the effect of a dielectric on the capacitance of a condenser.

Writing eqn. (3.6), viz. $Q_i = \sum c_{ij} V_j$, in matrix form $\mathbf{Q} = \mathbf{CV}$, we conclude that $\mathbf{V} = \mathbf{C}^{-1} \mathbf{Q}$ and $(p_{ij}) = \mathbf{C}^{-1}$. When the medium has dielectric constant K, the corresponding matrix is $K\mathbf{C}$ so that

$$(p'_{ij}) = (K\mathbf{C})^{-1} = K^{-1} \mathbf{C}^{-1} = (p_{ij}/K).$$

Hence the new coefficients of potential are $(\varepsilon_0/\varepsilon) p_{ij}$.

§4.3 DIELECTRICS

Example 2. Show that, if the electrostatic potential V depends only on the radial distance r, it satisfies

$$\frac{1}{r^2}\frac{d}{dr}\left(r^2\varepsilon_0 K\frac{dV}{dr}\right) = -\varrho,$$

where K is the dielectric constant and ϱ the charge density.

A spherical conducting core of total charge Z and radius a is surrounded by a gas whose density falls off with increasing r in such a way that its dielectric constant is given by $K(r) = 1 + \dfrac{b}{r}$ $(b > 0)$. Show that for $r > a$ the potential is given by

$$4\pi\varepsilon_0 V(r) = \frac{Z}{b}\ln\{K(r)\}.$$

Poisson's equation (4.14) when dielectric is present takes the form, in terms of spherical polar coordinates,

$$\text{div}\,(\varepsilon\,\text{grad}\,V)$$
$$= \frac{1}{r^2\sin\theta}\left\{\frac{\partial}{\partial r}\left(\varepsilon r^2\sin\theta\frac{\partial V}{\partial r}\right) + \frac{\partial}{\partial\theta}\left(\varepsilon\sin\theta\frac{\partial V}{\partial\theta}\right) + \frac{\partial}{\partial\psi}\left(\frac{\varepsilon}{\sin\theta}\frac{\partial V}{\partial\psi}\right)\right\} = -\varrho.$$

In the spherically symmetric case this equation reduces to

$$\frac{1}{r^2}\frac{d}{dr}\left(\varepsilon r^2\frac{dV}{dr}\right) = -\varrho, \tag{1}$$

which assumes the stated form when we write $\varepsilon = K\varepsilon_0$.

Since the core of radius a is a conductor carrying a charge Z, on the surface $r = a$ there is a surface density of charge $Z/(4\pi a^2)$. In the region $r \geq a$, $\varepsilon = \varepsilon_0(1+b/r)$. Hence, to determine the field we find a solution of eqn. (1), with $\varrho = 0$, which satisfies the following conditions

$$r = a; \quad \left(\varepsilon\frac{dV}{dr}\right)_{r=a} = -\frac{Z}{4\pi a^2}, \tag{2}$$

$$r \to \infty; \quad V \to 0. \tag{3}$$

On integrating eqn. (1) once we obtain $\varepsilon\,dV/dr = A/r^2$, and from condition (2) we find $-Z/(4\pi a^2) = A/a^2$, i.e. $A = -Z/(4\pi)$.

Therefore
$$\frac{dV}{dr} = -\frac{Z}{4\pi\varepsilon_0}\frac{1}{r(b+r)} = \frac{Z}{4\pi\varepsilon_0 b}\left\{\frac{1}{b+r} - \frac{1}{r}\right\}$$

and so
$$V = \frac{Z}{4\pi\varepsilon_0 b}\ln\left(1+\frac{b}{r}\right) + B.$$

Condition (3) now gives $B = 0$. Therefore

$$V(r) = \frac{Z}{4\pi\varepsilon_0 b}\ln\{K(r)\}.$$

Example 3. A given sphere of isotropic dielectric, of radius a, in which the dielectric constant $K(r)$ at distance r from the centre is a differentiable function of r only, is introduced into an electric field of uniform intensity E in free space. Prove that the electrostatic potential V at an internal point is given by an expression of the form

$$V = R(r)\cos\theta,$$

where R is a function of r only and the spherical polar coordinate θ is such that $\theta = 0$ gives the direction of the undisturbed external field.

124   ELEMENTARY ELECTROMAGNETIC THEORY

Show that R, K are related by the equation

$$\frac{d}{dr}\left(r^2 K \frac{dR}{dr}\right) = 2KR.$$

If $R = Ar + Br^2$, where A, B are constants, find a formula for K. Also determine the values of A, B in this case if $K(0) = 4$, $K(a) = 1$.

To solve this problem completely we must find V at all points of space both inside and outside the dielectric sphere $r = a$. The fact that there is no volume charge implies that

for $0 \leq r \leq a$, $\text{div}(\varepsilon \,\text{grad}\, V) = 0$, $\varepsilon = \varepsilon_0 K(r)$;   (1)

for $a \leq r$, $\varepsilon_0 \nabla^2 V = 0$.   (2)

There must also be continuity conditions at the surface $r = a$; they are

$$V_+ = V_-; \quad \left(\varepsilon_0 \frac{\partial V}{\partial n}\right)_+ - \left(\varepsilon \frac{\partial V}{\partial n}\right)_- = 0. \quad (3), (4)$$

Condition (4) arises because there is no "true" surface charge on the sphere. Finally, as $r \to \infty$, V must correspond to the uniform field, i.e.

$$V \simeq -Ez = -Er \cos\theta \quad \text{as} \quad r \to \infty. \quad (5)$$

At this point we use the "inspired guess" to which reference was made on p. 75. Since the field at infinity corresponds to the potential function (5) in which the variables are separated, we look for separable solutions of the partial differential equations of (1) and (2) in which the variation with θ is given by the factor $\cos\theta$. When (or if) we find a separable solution of this type which satisfies the conditions (1)–(5) (for a given K) the *uniqueness theorem* implies that we have obtained the only solution.

Substitution of the suggested, separated, form for V into eqn. (1) gives

$$\frac{\cos\theta}{r^2} \frac{d}{dr}\left(\varepsilon_0 K r^2 \frac{dR}{dr}\right) + \frac{\varepsilon_0 KR}{r^2 \sin\theta} \frac{d}{d\theta}(-\sin^2\theta) = 0,$$

i.e. $$\frac{1}{r^2}\frac{d}{dr}\left(Kr^2 \frac{dR}{dr}\right) - \frac{2KR}{r^2} = 0, \quad (6)$$

which is equivalent to the required differential equation.

When we are given that $R = Ar + Br^2$ we find from eqn. (6) that

$$\frac{dK}{dr} = -\frac{4BK}{A+2Br}, \quad \text{i.e.} \quad K = \frac{C}{(A+2Br)^2}, \quad (7)$$

where C is an arbitrary constant. Since $K > 0$, C must be positive and we may write $C = c^2$. The given boundary values for K lead to

$$4 = \frac{c^2}{A^2}, \quad 1 = \frac{c^2}{(A+2Ba)^2}.$$

Hence $A = \pm\tfrac{1}{2}c$, $A + 2Ba = \pm c$.

There are four possible values for A, B:

(a) $A = \tfrac{1}{2}c$, $B = \tfrac{1}{4}(c/a)$;   (b) $A = -\tfrac{1}{2}c$, $B = -\tfrac{1}{4}(c/a)$;
(c) $A = \tfrac{1}{2}c$, $B = -\tfrac{3}{4}(c/a)$;   (d) $A = -\tfrac{1}{2}c$, $B = \tfrac{3}{4}(c/a)$.

Corresponding to these values K must have one of the expressions

$$K = \frac{4a^2}{(r+a)^2} \quad \text{from (1) and (2); or}$$

$$K = \frac{4a^2}{(a-3r)^2} \quad \text{from (3) and (4).} \quad (8)$$

§ 4.3 DIELECTRICS 125

The corresponding functions $R(r)$ are:

(a) $R(r) = \dfrac{c}{2}\left(r+\dfrac{r^2}{2a}\right);$ (b) $R(r) = -\dfrac{c}{2}\left(r+\dfrac{r^2}{2a}\right);$

(c) $R(r) = \dfrac{c}{2}\left(r-\dfrac{3r^2}{2a}\right);$ (d) $R(r) = -\dfrac{c}{2}\left(r-\dfrac{3r^2}{2a}\right).$

We reject the solutions (c), (d) because they correspond to an infinite value for K at $r = a/3$.

Although the value of K given in eqn. (8) is determined as a function of r, the function $R(r)$ giving the potential is not determined because the value of c is arbitrary in (a) and (b). In order to find the unique solution we must now use the further information given for the region $r \geqslant a$.

The separable solution of (Laplace's) equation (2) is $V = f(r) \cos \theta$ where

$$\frac{1}{r^2}\frac{d}{dr}\left(r^2\frac{df}{dr}\right) - \frac{2f}{r^2} = 0, \quad \text{i.e.} \quad f(r) = \frac{A_1}{r^2} + B_1 r,$$

where A_1 and B_1 are constants. Condition (5) leads to $B_1 = -E$. Condition (3) gives

$$A_1/a^2 - Ea = Aa + Ba^2 = 3ca/4;$$

condition (4) gives, since $K = 1$ where $r = a$,

$$\varepsilon_0(-2A_1/a^3 - E) = \varepsilon_0(A + 2Ba) = \varepsilon_0 c.$$

Therefore $\quad \dfrac{A_1}{a^3} = \dfrac{E}{10}, \quad B_1 = -E, \quad c = -\dfrac{6E}{5}.$

These results apply to the expressions for $R(r)$ in case (a). For case (b) we obtain

$$\frac{A_1}{a^3} = \frac{E}{10}, \quad B_1 = E, \quad c = \frac{6E}{5}.$$

We thus obtain the unique solution for the problem as

$$A = \frac{3E}{5}, \quad B = -\frac{3Ea}{10}, \quad V = -\frac{3E}{5}\left(r+\frac{r^2}{2a}\right)\cos\theta, \quad 0 \leqslant r \leqslant a;$$

$$A_1 = \frac{Ea^3}{10}, \quad B_1 = -E; \quad V = E\left(\frac{a^3}{5r^2} - r\right)\cos\theta, \quad r \geqslant a;$$

and $\quad K = 4a^2/(r+a)^2.$

Example 4. A long cylinder of radius a, made of material of uniform dielectric constant K is placed with its axis at right-angles to a uniform electric field of strength E. Find the potential function giving the field at any point.

We adapt the method used in the previous example to find a separable solution. This arrangement gives a two-dimensional field, for which we use polar coordinates r, θ in the plane perpendicular to the axis of the cylinder. The potential function V has to satisfy the following conditions:

$$0 \leqslant r \leqslant a, \quad \text{div}\,(\varepsilon_0 K\,\text{grad}\,V) = 0; \tag{1}$$

$$r \geqslant a, \quad \varepsilon_0 \nabla^2 V = 0. \tag{2}$$

On the dividing surface $r = a$ the continuity conditions are

$$V_- = V_+, \quad \left(\varepsilon_0 K \frac{\partial V}{\partial v}\right)_- = \left(\varepsilon_0 \frac{\partial V}{\partial v}\right)_+. \tag{3, 4}$$

At infinity the field must be the uniform field E, i.e.
$$V \approx -Er \cos \theta, \quad r \to \infty. \tag{5}$$
Since K is uniform both eqns. (1) and (2) reduce to the two-dimensional form of Laplace's equation, which is
$$\frac{\partial^2 V}{\partial r^2} + \frac{1}{r}\frac{\partial V}{\partial r} + \frac{1}{r^2}\frac{\partial^2 V}{\partial \theta^2} = 0.$$
We seek separable solutions of the form $V = f(r) \cos \theta$. Then
$$\cos \theta \left\{ \frac{d^2 f}{dr^2} + \frac{1}{r} \frac{df}{dr} \right\} - \frac{f(r) \cos \theta}{r^2} = 0.$$
Therefore
$$r^2 \frac{d^2 f}{dr^2} + r \frac{df}{dr} - f = 0; \quad f = \frac{A}{r} + Br.$$
We can write
$$0 \leqslant r \leqslant a, \quad V = \left(\frac{A_1}{r} + B_1 r \right) \cos \theta,$$
$$r \geqslant a, \quad V = \left(\frac{A_2}{r} + B_2 r \right) \cos \theta.$$

Since $r = 0$ is included in the region of the dielectric, we must put $A_1 = 0$, so that V remains finite where $r = 0$.

Condition (5) shows that $B_2 = -E$.

Condition (3) gives $B_1 a = \dfrac{A_2}{a} - Ea$.

Condition (4) gives $KB_1 = -\dfrac{A_2}{a^2} + B_2$.

These equations have solutions
$$A_1 = 0, \quad B_1 = -\frac{2E}{K+1}, \quad A_2 = \frac{K-1}{K+1} Ea^2, \quad B_2 = -E.$$
Therefore
$$\text{for } 0 \leqslant r \leqslant a, \quad V = -\frac{2Er \cos \theta}{K+1};$$
$$\text{for } r \geqslant a, \quad V = E \left(\frac{K-1}{K+1} \frac{a^2}{r} - r \right) \cos \theta.$$

This shows that inside the dielectric the field is of uniform strength $2E/(K+1)$ and is parallel to the original field.

The polarization is uniform and given by $\mathbf{P} = \mathbf{D} - \varepsilon_0 \mathbf{E}$.

Therefore
$$|\mathbf{P}| = \frac{2\varepsilon_0 KE}{K+1} - \frac{2\varepsilon_0 E}{K+1} = \varepsilon_0 \frac{2(K-1)E}{K+1}.$$

For points outside the dielectric the field is the uniform field augmented by the field due to a line of dipoles along the axis of moment per unit length (of cylinder)
$$\mu = 2\pi\varepsilon_0 \left(\frac{K-1}{K+1} \right) Ea^2 = \pi a^2 |\mathbf{P}|.$$

The solutions of examples 3 and 4 are important results and in fact the method used is the general method given in C.M., vol. 4, § 2.1(3) for example 3, and C.M., vol. 4, § 2.1(1) for example 4. The function giving the angular dependence in example 3 is the Legendre function $P_1 (\cos \theta) = \cos \theta$. Further examples are given in Vol. 2, Chap. 8.

---

§ 4.3 DIELECTRICS

Example 5. Two conductors A and B, of which B surrounds A, are maintained at potentials V_A and V_B. The region between A and a surface S on which the potential initially has an intermediate constant value V_0 is now filled with dielectric of dielectric constant K. Find the new potential on the outer surface of the dielectric and show that the potential, at any point occupied by the dielectric, is altered from V to V', where

$$V' = \frac{V_A(V_0-V_B)(K-1) + V(V_A-V_B)}{K(V_0-V_B) + (V_A-V_0)}.$$

Suppose that V is the potential function satisfying $\nabla^2 V = 0$ and which assumes the value V_A, V_B on the inner and outer surfaces, and the value V_0 on the intermediate surface S

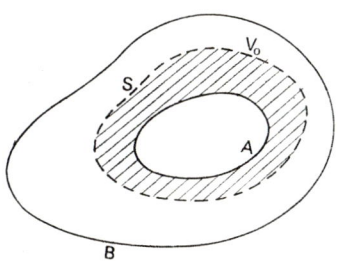

FIG. 4.2

in the absence of the dielectric, Fig. 4.2. Since V_A, V_B are maintained at the same constant values after the introduction of the dielectric, we try to find a solution in which the potential function is

$$\phi = A_1 + \lambda_1 V \quad \text{inside the intermediate surface,}$$
$$ A_2 + \lambda_2 V \quad \text{outside the intermediate surface,}$$

where A_1, A_2, λ_1, λ_2 are constants. Therefore ϕ satisfies eqn. (4.14) at all points between A and B.

Inside S, $\operatorname{div}(\varepsilon_0 K \operatorname{\mathbf{grad}} \phi) = \varepsilon_0 K \nabla^2 \phi = \varepsilon_0 K \nabla^2 (A_1 + \lambda_1 V) = \varepsilon_0 K \lambda_1 \nabla^2 V = 0$.

Outside S, $\quad \varepsilon_0 \nabla^2 \phi = \varepsilon_0 \nabla^2 (A_2 + \lambda_2 V) = \varepsilon_0 \lambda_2 \nabla^2 V = 0$.

At all points on A,

$$\phi = A_1 + \lambda_1 V_A = V_A, \tag{1}$$

and on B,

$$\phi = A_2 + \lambda_2 V_B = V_B. \tag{2}$$

The function ϕ must be continuous across S and so

$$A_1 + \lambda_1 V_0 = A_2 + \lambda_2 V_0 \tag{3}$$

for all points on S.

The continuity of of $(\mathbf{D}.\hat{\mathbf{v}})$ across the boundary S gives

$$\varepsilon_0 K \frac{\partial}{\partial v}(A_1 + \lambda_1 V) = \varepsilon_0 \frac{\partial}{\partial v}(A_2 + \lambda_2 V).$$

Therefore

$$\varepsilon_0 K \lambda_1 \frac{\partial V}{\partial v} = \varepsilon_0 \lambda_2 \frac{\partial V}{\partial v}$$

and so

$$\lambda_2 = K\lambda_1. \tag{4}$$

128   ELEMENTARY ELECTROMAGNETIC THEORY

The solution of the four equations (1)–(4) for the four constants A_1, A_2, λ_1, λ_2 shows that

$$\phi = \frac{V_A(K-1)(V_0-V_B)+V(V_A-V_B)}{K(V_0-V_B)+V_A-V_0} \tag{5}$$

and that, on the surface S, $\phi = V_0'$, where

$$V_0' = \frac{V_0(KV_A-V_B)-(K-1)V_AV_B}{K(V_0-V_B)+(V_A-V_0)}.$$

Since the function ϕ, obtained in eqn. (5), satisfies all the required conditions, viz. $\phi = V_A$, V_B on the outer and inner surfaces, div $(\varepsilon_0 K \text{ grad } \phi) = 0$ at all points, and satisfies the usual boundary conditions at the surface S of discontinuity in K, it must give the field because of the uniqueness theorem.

The essence of the solution is in the assumption that ϕ is a linear function of V; the remainder of the working is the determination of the coefficients in this linear relation.

Example 6. The region between two concentric spherical metal shells is filled on one side of a diametral plane with a uniform dielectric of dielectric constant K_1, and on the other side with a uniform dielectric of dielectric constant K_2. The shells are insulated; the inner, of radius a, carries a total charge q_1, and the outer, of radius b, carries a total charge q_2. Find the electric potential and field at all points, showing explicitly that they satisfy the boundary conditions.

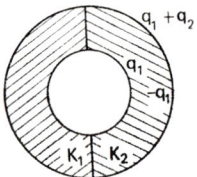

FIG. 4.3

The total charges on the inner and outer surfaces of the shells are shown in Fig. 4.3. For $r > b$,

$$D_r = \frac{q_1+q_2}{4\pi r^2}, \qquad V = \frac{q_1+q_2}{4\pi\varepsilon_0 r}. \tag{1}$$

Between the shells, i.e. for $a < r < b$, we *try*

$$V = A + \frac{B}{r}, \tag{2}$$

where A, B are constants to be determined. This expression for V satisfies Laplace's equation, has no singularities for $a < r < b$, makes the shells equipotentials and across the dielectric interface $V_+ - V_- = 0$, $(\partial V/\partial n)_+ = 0 = (\partial V/\partial n)_-$ so that the interface is uncharged. It follows, by the uniqueness theorem, that if we choose A, B so that the charge on the inner shell is q_1 and the charge on the inner surface of the outer shell is $-q_1$, we have obtained the unique solution of the prescribed problem. Using eqn. (1), the potential condition at the outer shell gives

$$A + \frac{B}{b} = \frac{q_1+q_2}{4\pi\varepsilon_0 b}. \tag{3}$$

§ 4.3 DIELECTRICS

For $a < r < b$,

$$E_r = -\frac{dV}{dr} = \frac{B}{r^2}, \quad D_r = \frac{\varepsilon_0 KB}{r^2}.$$

It follows from Gauss's theorem, that

$$\oint_{r=a} D_r \, dS = 2\pi B \varepsilon_0 (K_1 + K_2) = q_1.$$

Therefore $\quad B = \dfrac{q_1}{2\pi\varepsilon_0(K_1+K_2)}, \quad A = \dfrac{q_2}{4\pi\varepsilon_0 b} + \dfrac{q_1}{4\pi\varepsilon_0 b}\left(1 - \dfrac{2}{K_1+K_2}\right).$

Note that the relation $D_n = \sigma$ gives the surface densities of charge as follows:

$\dfrac{K_1 q_1}{2\pi(K_1+K_2)a^2}$ on the left-hand half of the inner shell,

$\dfrac{K_2 q_1}{2\pi(K_1+K_2)a^2}$ on the right-hand half of the inner shell,

$\dfrac{-K_1 q_1}{2\pi(K_1+K_2)b^2}$ on the left-hand half of the inner surface of the outer shell,

$\dfrac{-K_2 q_1}{2\pi(K_1+K_2)b^2}$ on the right-hand half of the inner surface of the outer shell,

$\dfrac{q_1 + q_2}{4\pi b^2}$ on the outer surface of the outer shell.

Exercises 4.3

1. From the divergence theorem $\iiint \text{div } \mathbf{A} \, d\tau = \oiint \mathbf{A} \cdot d\mathbf{S}$ prove Kelvin's generalization of Green's theorem by adopting a special form for $\mathbf{A}$, viz. $\mathbf{A} = \varepsilon U \text{ grad } U$, where $\varepsilon > 0$.
2. Write out the proof of the uniqueness theorem (p. 121).
3. Write out the proof of Green's reciprocal theorem (p. 121).
4. A conducting sphere of radius a is surrounded by a concentric conducting sphere of radius $4a$. The region $a < r < 2a$ is filled with matter of dielectric constant K_1; the region $2a < r < 4a$ with matter of dielectric constant K_2. The outer sphere carries a charge Q, and the inner is earthed. Find the electric field through all space, and show that the charge on the inner sphere is

$$-\frac{Q}{\left(1+\dfrac{1}{K_2}+\dfrac{2}{K_1}\right)}.$$

5. Two concentric conducting spheres of radii a, b ($b > a$) carry charges q_1 and q_2 respectively. The space between the spheres is filled with a substance of dielectric constant $K = 1/\{1-(\lambda r^2/b^2)\}$, where λ is a constant ($0 < \lambda < 1$), at a distance r from the centre of the spheres. Deduce the density of charge on each of the four surfaces of the spheres, find the potential at any point between the spheres, and prove that the potential of the inner sphere is

$$\frac{1}{4\pi\varepsilon_0}\left\{\frac{q_1}{a}+\frac{q_2}{b}-\lambda q_1\frac{(b-a)}{b^2}\right\},$$

6. The space between two concentric conducting spheres is filled on one side of a diametral plane with dielectric of specific inductive capacity K and on the other side with dielectric of specific inductive capacity K'. The inner sphere is of radius a and has a charge q. Show that the force on it perpendicular to this diametral plane is

$$\frac{\varepsilon_0(K-K')q^2}{8\pi(K+K')^2 a^2}.$$

7. An infinitely long cylindrical shell of external and internal radii r_1 and r_2 respectively is of material of dielectric constant K. It is placed, uncharged, in a uniform electric field F perpendicular to its axis. Prove that the field within the region $0 \leqslant r \leqslant r_2$ is reduced by a factor

$$1 \bigg/ \left\{1 + \frac{(K-1)^2}{4K}\left(1 - \frac{r_2^2}{r_1^2}\right)\right\}.$$

8. A uniform dielectric shell is bounded internally and externally by concentric spheres of radii λa and a respectively. It is placed in a uniform electric field of intensity E. If K is the dielectric constant, show that the field inside the central cavity is uniform and of intensity

$$9KE/\{9K + 2(1-\lambda^3)(K-1)^2\}.$$

4.4 The energy of the electrostatic field

We consider an electrostatic field of which the sources are: volume distribution ϱ, surface distribution σ on a surface K, charges Q_i on conductors S_i, and point charges Q_l (Fig. 4.4). We augment these charges by arbitrary

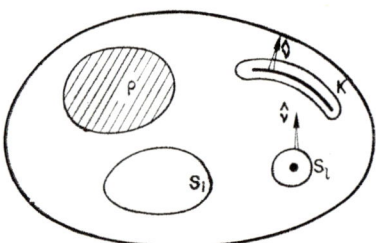

Fig. 4.4

amounts $\delta\varrho$, $\delta\sigma$, δQ_i, δQ_l. In order to place these additional charges in position the work which must be done is

$$\delta W = \iiint V \delta\varrho \, d\tau + \iint_K V \delta\sigma \, dS + \sum_i V_i \delta Q_i + \sum_l V_l \delta Q_l. \qquad (4.19)$$

This expression is correct to the first order. We suppose that this augmentation of the sources of the field causes changes $\delta \mathbf{D}$, $\delta \mathbf{E}$ in the field vectors. (These additional charges also cause an increase δV in the potential at any point but the extra work required to bring the additional charges into posi-

§ 4.4 DIELECTRICS 131

tion because of this change of potential is of the second order, e.g. $\delta Q_i\, \delta V_i$. Hence the above expression is correct to the first order.) The change δD must satisfy the following conditions

$$\operatorname{div} \delta \boldsymbol{D} = \delta\varrho, \quad \hat{\boldsymbol{v}} \cdot (\delta \boldsymbol{D}_+ - \delta \boldsymbol{D}_-) = \delta\sigma, \quad \iint_{S_i} \hat{\boldsymbol{v}} \cdot \delta \boldsymbol{D}\, dS = \delta Q_i,$$

$$\iint_{S_l} \hat{\boldsymbol{v}} \cdot \delta \boldsymbol{D}\, dS = \delta Q_l. \tag{4.20}$$

The surface S_0 encloses all charges in the field; S_K is a closed surface enveloping any surface K carrying surface charges σ, and S_l are small spheres enclosing point charges Q_l; S_i are conducting surfaces. The vector $\hat{\boldsymbol{v}}$ is directed out of each surface. We transform eqn. (4.19) to a form closely similar to eqn. (3.32) by applying the divergence theorem to the volume exterior to S_K, S_i, S_l and inside S_0 (see Fig. 4.4).

Using the first of eqns. (4.20) we transform the integral

$$\iiint V\delta\varrho\, d\tau = \iiint V \operatorname{div} \delta \boldsymbol{D}\, d\tau = \iiint \{\operatorname{div}(V\delta \boldsymbol{D}) - \delta \boldsymbol{D} \cdot \operatorname{grad} V\}\, d\tau$$

$$= \oiint V(\delta \boldsymbol{D} \cdot \hat{\boldsymbol{n}})\, dS + \iiint \boldsymbol{E} \cdot \delta \boldsymbol{D}\, d\tau, \tag{4.21}$$

where the surface integral is taken over all the boundary surfaces of the volume of integration (the unit normal out of the volume of integration being $\hat{\boldsymbol{n}}$), and the volume integral is taken through the volume described above. As in § 3.6 we note the following, bearing in mind that $\hat{\boldsymbol{n}}$ in eqn. (4.21) is drawn out of the volume of integration.

(1) $\oiint_{S_0} V(\delta \boldsymbol{D} \cdot \hat{\boldsymbol{n}})\, dS \to 0$ as S_0 goes to infinity because

$$V = O(1/R), \quad |\delta \boldsymbol{D}| = O(1/R^2) \text{ as } R \to \infty.$$

(2) On the conductor S_i,

$$\oiint_{S_i} V(\delta \boldsymbol{D} \cdot \hat{\boldsymbol{n}})\, dS = -V_i \oiint_{S_i} \delta \boldsymbol{D} \cdot \hat{\boldsymbol{v}}\, dS = -V_i \delta Q_i$$

remembering that $\hat{\boldsymbol{v}}$ is the usual outward drawn normal to S_i.

(3) On the envelope $\oiint_{S_K} V(\delta \boldsymbol{D} \cdot \hat{\boldsymbol{n}})\, dS$ tends to a limit as S_K tends to coincide with K.

$$\oiint_{S_K} V(\delta \boldsymbol{D} \cdot \hat{\boldsymbol{n}})\, dS \to \iint_K V\hat{\boldsymbol{v}} \cdot (-\delta \boldsymbol{D}_+ + \delta \boldsymbol{D}_-)\, dS = -\iint_K V\delta\sigma\, dS.$$

(4) On the small sphere S_l, the direction of $\hat{n}$ in eqn. (4.21) is into S_l, i.e. $\hat{v} = -\hat{n}$. Hence on this small sphere $V \approx V_l + O(\eta)$ so that, as the radius η of the sphere tends to zero,

$$\oint_{S_l} V(\delta \mathbf{D} \cdot \hat{n}) \, dS \rightarrow -V_l \oint_{S_l} (\delta \mathbf{D} \cdot \hat{v}) \, dS = -V_l \delta Q_l,$$

where we have used Gauss's theorem.

Collecting results (1)–(4), we conclude that

$$\iiint V \delta \varrho \, d\tau = \iiint \mathbf{E} \cdot \delta \mathbf{D} \, d\tau - \iint_K V \delta \sigma \, dS - \sum_i V_i \delta Q_i - \sum_l V_l \delta Q_l. \tag{4.22}$$

Therefore
$$\delta W = \iiint (\mathbf{E} \cdot \delta \mathbf{D}) \, d\tau. \tag{4.23}$$

We emphasize that eqn. (4.23) is only correct to the first order in the increments.

For an isotropic, linear medium $\delta \mathbf{D} = \varepsilon \delta \mathbf{E}$, since ε is independent of the field, and so

$$\mathbf{E} \cdot \delta \mathbf{D} = \varepsilon \mathbf{E} \cdot \delta \mathbf{E} = \tfrac{1}{2} \varepsilon \delta(E^2) = \delta(\tfrac{1}{2} \mathbf{E} \cdot \mathbf{D}).$$

Therefore
$$\delta W = \delta \left\{ \tfrac{1}{2} \iiint \mathbf{E} \cdot \mathbf{D} \, d\tau \right\}.$$

Since $W = 0$ when there is no field, we may integrate this relation to obtain

$$W = \tfrac{1}{2} \iiint \mathbf{E} \cdot \mathbf{D} \, d\tau = \tfrac{1}{2} \iiint \varepsilon E^2 \, d\tau. \tag{4.24}$$

In addition to making the assumption of a linear isotropic medium we have interchanged two processes of integration. In eqn. (4.23) the integration over the space variables is used to evaluate the increment δW; to obtain eqn. (4.24) we "integrated" the increment $\mathbf{E} \cdot \delta \mathbf{E}$ to obtain $\delta(\tfrac{1}{2} E^2)$, and then used $\tfrac{1}{2} \varepsilon E^2 = \tfrac{1}{2} \mathbf{E} \cdot \mathbf{D}$ as the integrand for the space variables.

Following the suggestion in § 3.6 we interpret result (4.24) as implying an energy density in the field of amount $\tfrac{1}{2} \mathbf{E} \cdot \mathbf{D}$. Since $\mathbf{D} = \varepsilon_0 \mathbf{E} + \mathbf{P}$,

$$\tfrac{1}{2} \mathbf{E} \cdot \mathbf{D} = \tfrac{1}{2} \mathbf{E} \cdot (\varepsilon_0 \mathbf{E} + \mathbf{P}) = \tfrac{1}{2} \varepsilon_0 E^2 + \tfrac{1}{2} \mathbf{E} \cdot \mathbf{P}. \tag{4.25}$$

We interpret the last two terms of eqn. (4.25) as follows: $\tfrac{1}{2} \varepsilon_0 E^2$ is the energy required to "fill the space" with a field $\mathbf{E}$; the term $\tfrac{1}{2} \mathbf{E} \cdot \mathbf{P}$ is the additional energy required to separate the charges in the medium against the forces holding them in position, i.e. to produce the polarization $\mathbf{P}$ (cf. the energy $\tfrac{1}{2} Tx$ in a stretched spring), and represents the energy of "strain" in the material of

§ 4.4 DIELECTRICS

the dielectric. Once again we remind the reader that, since we are now considering representations of material bodies, other sources of energy, such as thermal energy, may be involved. Thus, the "strain" of the dielectric corresponding to the polarization may take place at the expense of thermal energy and the temperature of the material would fall. Such matters should strictly be investigated by the methods of thermodynamics; however, if we ensure, by the supply of heat if necessary, that any changes in the state of the system take place slowly at a fixed temperature, then the energies considered here are changes in the "available energy" (or free energy) of the system. Similarly any displacements of position which are assumed to occur in evaluating ponderomotive forces must take place while the temperature of every part of the system remains fixed.

Example 1. An insulated spherical conductor in air carries a charge q. If the conductor is now surrounded by a concentric spherical shell of dielectric of radii b and c, $(c > b)$, whose specific inductive capacity is a function $k(r)$ of the radial distance r from the centre, prove that the loss of energy is

$$\frac{q^2}{8\pi\varepsilon_0} \int_b^c \left\{1 - \frac{1}{k(r)}\right\} \frac{dr}{r^2}.$$

In both cases

$$D_r = \frac{q}{4\pi r^2}.$$

In the first case (air)

$$V = \frac{q}{4\pi\varepsilon_0 r}, \qquad W_1 = \frac{q^2}{8\pi\varepsilon_0} \int_a^\infty \frac{dr}{r^2}.$$

Here a is the radius of the sphere.

In the second case, when the dielectric shell is placed in position,

$$E_r = \begin{cases} \dfrac{q}{4\pi\varepsilon_0 r^2} & \text{for } a \leqslant r \leqslant b, \\ \dfrac{q}{4\pi\varepsilon_0 k(r) r^2} & \text{for } b \leqslant r \leqslant c, \\ \dfrac{q}{4\pi\varepsilon_0 r^2} & \text{for } c \leqslant r. \end{cases}$$

Therefore

$$W_2 = \frac{1}{2} \iiint \mathbf{D} \cdot \mathbf{E} \, d\tau$$

$$= \frac{q^2}{16\pi^2} \int_a^b \frac{4\pi r^2 \, dr}{\varepsilon_0 r^4} + \frac{q^2}{16\pi^2} \int_b^c \frac{4\pi r^2 \, dr}{\varepsilon_0 k(r) r^4} + \frac{q^2}{16\pi^2} \int_c^\infty \frac{4\pi r^2 \, dr}{\varepsilon_0 r^4}$$

$$= \frac{q^2}{8\pi\varepsilon_0} \left\{ \int_a^b \frac{dr}{r^2} + \int_b^c \frac{dr}{k(r) r^2} + \int_c^\infty \frac{dr}{r^2} \right\}.$$

Hence
$$W_1 - W_2 = \frac{q^2}{8\pi\varepsilon_0}\left\{\int_a^b \frac{dr}{r^2} + \int_b^c \frac{dr}{k(r)r^2} + \int_c^\infty \frac{dr}{r^2}\right\}$$
$$- \frac{q^2}{8\pi\varepsilon_0}\left\{\int_a^b \frac{dr}{r^2} + \int_b^c \frac{dr}{r^2} + \int_c^\infty \frac{dr}{r^2}\right\}$$
$$= \frac{q^2}{8\pi\varepsilon_0}\int_b^c \left\{1 - \frac{1}{k(r)}\right\}\frac{dr}{r^2}.$$

The fact that the electrostatic energy decreases illustrates the result of example 2 below.

Example 2. Prove that if, while the total charge on each conductor of an electrostatic system is held constant, there is an increase δK in the dielectric constant of the intervening medium, then the electrostatic energy of the system is decreased by an amount

$$\frac{\varepsilon_0}{2}\int E^2 \delta K \, d\tau,$$

where the integral is over the volume between the conductors.

Let E, D denote the field quantities before the change of dielectric constant, and E', D' the corresponding quantities after the change.

Each field corresponds to the same total charge on each conductor, and to identical volume, surface and point charges (if there are any such charges).

Therefore

$$\text{div } \boldsymbol{D} = \varrho = \text{div } \boldsymbol{D}', \quad \hat{\boldsymbol{v}}\cdot(\boldsymbol{D}_+ - \boldsymbol{D}_-) = \sigma = \hat{\boldsymbol{v}}\cdot(\boldsymbol{D}'_+ - \boldsymbol{D}'_-),$$

$$\iint_{S_i} \boldsymbol{D}\cdot\hat{\boldsymbol{v}} \, dS = Q_i = \iint_{S_i} \boldsymbol{D}'\cdot\hat{\boldsymbol{v}} \, dS, \quad \boldsymbol{D} = \varepsilon_0 K \boldsymbol{E}, \quad \boldsymbol{D}' = \varepsilon_0(K + \delta K)\boldsymbol{E}',$$

and both fields have identical singularities at point charges Q_i. If $\boldsymbol{E} = -\text{grad } V$, $\boldsymbol{E}' = -\text{grad } V'$, both V and V' are constant on each conductor, but in general $V_i \neq V'_i$.

Now

$$W = \frac{1}{2}\iiint V\varrho \, d\tau + \frac{1}{2}\iint V\sigma \, dS + \frac{1}{2}\sum_i Q_i V_i + \frac{1}{2}\sum_i Q_i V_i$$

with a similar expression for W'. Therefore

$$W' - W = \frac{1}{2}\iiint (V' - V)\varrho \, d\tau + \frac{1}{2}\iint (V' - V)\sigma \, dS + \frac{1}{2}\sum (V'_i - V_i)Q_i$$
$$+ \frac{1}{2}\sum (V'_i - V_i)Q_i$$

since the charges are not altered in the process. But the alternative expression for the energy is

$$W' = \frac{\varepsilon_0}{2}\iiint (K + \delta K)E'^2 \, d\tau, \quad W = \frac{\varepsilon_0}{2}\iiint KE^2 \, d\tau.$$

Therefore
$$W' - W = \frac{\varepsilon_0}{2}\iiint \delta K E^2 \, d\tau + \varepsilon_0 \iiint K\boldsymbol{E}\cdot(\boldsymbol{E}' - \boldsymbol{E}) \, d\tau,$$

where we have ignored the second-order difference between

$$E'^2 \delta K \quad \text{and} \quad E^2 \delta K.$$

§ 4.4 DIELECTRICS

Now

$$\varepsilon_0 \iiint K\mathbf{E}\cdot(\mathbf{E}'-\mathbf{E})\,d\tau = -\iiint \mathbf{D}\cdot\mathrm{grad}\,(V'-V)\,d\tau$$

$$= \iiint (V'-V)\,\mathrm{div}\,\mathbf{D}\,d\tau - \iiint \mathrm{div}\,[(V'-V)\mathbf{D}]\,d\tau.$$

We exclude surface charges σ, and point charges Q_l from the volume of integration by suitable surfaces as in § 2.9 and we enclose the whole in a large surface S_0. In the limiting case the second integral becomes

$$-\varepsilon_0 \iiint \mathrm{div}\,[(V'-V)\mathbf{D}]\,d\tau = -\varepsilon_0 \iint (V'-V)\mathbf{D}\cdot\hat{\mathbf{n}}\,dS,$$

where $\hat{\mathbf{n}}$ is the unit normal drawn *out* of the above volume of integration. In the limiting case the contributions from these surface integrals become:

over S_0, $\qquad \lim \iint (V'-V)\mathbf{D}\cdot\hat{\mathbf{n}}\,dS = 0;$

over surface charges, $\qquad \lim \iint_K (V'-V)\mathbf{D}\cdot\hat{\mathbf{n}}\,dS = -\iint (V'-V)\sigma\,dS;$

at point charges, $\qquad \lim \iint (V'-V)\mathbf{D}\cdot\hat{\mathbf{n}}\,dS = -(V_i'-V_i)Q_l;$

on conducting boundaries S_i, $\iint (V'-V)\mathbf{D}\cdot\hat{\mathbf{n}}\,dS = -(V_i'-V_i)Q_i.$

Therefore $\varepsilon_0 \iiint K\mathbf{E}\cdot(\mathbf{E}'-\mathbf{E})\,d\tau = \iiint (V'-V)\varrho\,d\tau + \iint_K (V'-V)\sigma\,dS$

$$+ \sum_i (V'-V_i)Q_i + \sum_l (V_l'-V_l)Q_l$$

$$= 2(W'-W)$$

Therefore $\qquad W'-W = \dfrac{\varepsilon_0}{2}\iiint E^2\delta K\,d\tau + 2(W'-W),$

i.e. $\qquad W'-W = -\dfrac{\varepsilon_0}{2}\iiint E^2\delta K\,d\tau.$

It follows from this result that a piece of dielectric experiences a force tending to draw it into regions of stronger field. The next example is an illustration of this in a special case.

Example 3. A parallel plate condenser consists of rectangular plates each of total area A and at separation a between which a parallel slab of dielectric of constant K and thickness t can be introduced. Neglecting edge effects calculate the variation in electrostatic energy as the slab is introduced:

(a) when the plates of the condenser are held at a constant potential difference V;
(b) when they have constant charges Q and $-Q$.

In each case what is the force on the slab when it covers a fraction x of the area of the plates?

We consider the condenser when the dielectric overlaps a fraction x of the plates, as shown in Fig. 4.5.

When the dielectric is inserted into the condenser in this manner the electrostatic energy is decreased (when the charges remain constant) so that there is a force tending to draw the dielectric into the field, i.e. to reduce the electrostatic energy. We calculate the energy in

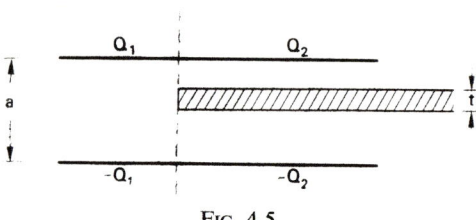

Fig. 4.5

terms of x and use the method of conservation energy (cf. § 3.4) to calculate the force on the slab.

Because we neglect all edge effects, the displacement D is uniform in a line normal to the plates, and we write down

$$D_1 = \sigma_1 = \frac{Q_1}{A(1-x)}, \qquad D_2 = \sigma_2 = \frac{Q_2}{Ax}.$$

Therefore $\quad E_1 = \dfrac{Q_1}{\varepsilon_0 A(1-x)}, \qquad E_2 = \dfrac{Q_2}{\varepsilon_0 Ax} \quad$ outside the dielectric,

$$E_2 = \frac{Q_2}{\varepsilon_0 K A x} \quad \text{inside the dielectric.}$$

Therefore $\quad V = \dfrac{Q_1 a}{\varepsilon_0 A(1-x)}, \qquad V = \dfrac{Q_2(a-t)}{\varepsilon_0 Ax} + \dfrac{Q_2 t}{\varepsilon_0 K A x}$

$$= \frac{Q_2}{\varepsilon_0 A x}\left[a - \left(1 - \frac{1}{K}\right)t\right]. \tag{1}$$

(a) When the plates are kept at constant potential difference V, by a battery, as x varies Q_1 and Q_2 must vary to keep V fixed. Hence, in general the battery adds charge to the system at a potential V and so supplies energy $V(\delta Q_1 + \delta Q_2)$ when x increases by the small amount δx.

The equation of conservation of energy is

$$V(\delta Q_1 + \delta Q_2) - F\delta x = \delta W,$$

where F is the force on the dielectric tending to increase x, and δW is the increment in the energy. Since V is constant we see that

$$\delta Q_1 = -\varepsilon_0 \frac{AV\delta x}{a}, \qquad \delta Q_2 = \frac{\varepsilon_0 A V \delta x}{a - (1 - 1/K)t}.$$

We calculate W in terms of the potential, thus

$$W = \frac{1}{2} V(Q_1 + Q_2) = \frac{1}{2}\varepsilon_0 A V^2 \left\{ \frac{1-x}{a} + \frac{x}{a-(1-1/K)t}\right\}.$$

Therefore $\quad \delta W = \dfrac{1}{2}\varepsilon_0 A V^2 \delta x \left\{ -\dfrac{1}{a} + \dfrac{1}{a-(1-1/K)t}\right\}.$

§ 4.5 DIELECTRICS 137

The equation of conservation of energy gives

$$\varepsilon_0 A V^2 \delta x \left\{ -\frac{1}{a} + \frac{1}{a-(1-1/K)t} \right\} - F\delta x = \tfrac{1}{2}\varepsilon_0 A V^2 \delta x \left\{ -\frac{1}{a} + \frac{1}{a-(1-1/K)t} \right\}.$$

Therefore
$$F = \frac{\varepsilon_0 A V^2}{2} \left\{ \frac{1}{a-(1-1/K)t} - \frac{1}{a} \right\}.$$

(b) When the charges are kept constant, the potential V alters while the total charge Q is redistributed between Q_1 and Q_2 as x increases. In this case, since no energy is supplied from a battery, conservation of energy gives

$$\delta W + F \delta x = 0.$$

From eqn. (1) we can write

$$\frac{Q_1}{(1-x)\{a-(1-1/K)t\}} = \frac{Q_2}{xa} = \frac{Q}{a-(1-x)(1-1/K)t},$$

where $Q = Q_1 + Q_2$. Therefore

$$V = \frac{Q_1 a}{\varepsilon_0 A(1-x)} = \frac{Qa\{a-(1-1/K)t\}}{\varepsilon_0 A\{a-(1-x)(1-1/K)t\}} \tag{2}$$

The energy is $\tfrac{1}{2}QV$ and so

$$W = \frac{Q^2 a\{a-(1-1/K)t\}}{2\varepsilon_0 A\{a-(1-x)(1-1/K)t\}}.$$

Therefore
$$\delta W = -F\delta x = -\frac{Q^2 a\{a-(1-1/K)t\}}{2\varepsilon_0 A\{a-(1-x)(1-1/K)t\}^2} \delta x (1-1/K)t.$$

Therefore
$$F = \frac{Q^2 a\{a-(1-1/K)t\}\,(1-1/K)t}{2\varepsilon_0 A\{a-(1-x)(1-1/K)t\}^2}.$$

By the use of eqn. (2) we eliminate Q and obtain

$$F = \frac{\varepsilon_0 A V^2 (1-1/K)t}{2a\{a-(1-1/K)t\}} = \frac{\varepsilon_0 A V^2}{2} \left\{ \frac{1}{a-(1-1/K)t} - \frac{1}{a} \right\},$$

which is the same expression as we obtained above (cf. § 3.4).

4.5 Minimum energy theorems

The energy of the electrostatic field, calculated in § 3.4. and § 4.4, is regarded as potential energy of the system consisting of charges, dielectrics, conductors, etc. When charges of opposite sign are allowed to neutralize each other and unbalanced charges are removed to infinite dispersal so that the field is reduced to zero, this electrostatic energy is returned in the work required for this process of dispersal. A general theorem of mechanics states that a system free to move under forces and constraints takes up a position of equilibrium for which the potential energy is stationary—a minimum when the equilibrium is stable. A similar theorem, known as *Kelvin's minimum energy theorem*, can be proved for the electrostatic field. However, in stating these theorems we must be quite clear as to what variations from the equilibrium position are contemplated.

For example, suppose a particle is free to move in a uniform gravitational field of infinite extent. When projected the particle never reaches an equilibrium state but will continue to move off to infinity (in the parabolic path of a projectile). The particle will reach equilibrium only if it is constrained in some way; for example it may be constrained to slide on a curve. Then the particle has equilibrium positions at the minima of the curve. (In fact, unless there is friction which dissipates energy, the particle will oscillate about such a position.)

Similarly, if a number of electric charges are left completely free, either they neutralize one another, or they repel one another off to infinity under their mutual repulsion. But if the charges are constrained in some way, then, by a suitable rearrangement subject to these constraints, there exists an equilibrium arrangement in which the electrostatic energy of the system is a minimum. The constraints we consider are these:

(1) All volume charge ϱ and point charges remain fixed.
(2) All surface charge σ, not situated on a conductor remains fixed, on surfaces S_L.
(3) All charge situated on a conductor is free to move on the surface of that conductor, S_i.
(4) Any polarizable medium is linear, and the fields satisfy the standard boundary conditions at infinity.

These conditions are embodied in the results

$$\text{div } \boldsymbol{D} = \varrho, \quad \boldsymbol{\hat{v}} \cdot (\boldsymbol{D}_+ - \boldsymbol{D}_-) = \sigma,$$

$$\iint \boldsymbol{D} \cdot \boldsymbol{\hat{v}} \, dS = Q_i$$

on conducting boundaries,

$$\boldsymbol{D} = \varepsilon \boldsymbol{E} = \varepsilon_0 \boldsymbol{E} + \boldsymbol{P}.$$

To prove Kelvin's theorem we consider a field, different from the original but which satisfies the above conditions. The new electric field is $\boldsymbol{E}+\boldsymbol{e}$, and the displacement vector is $\boldsymbol{D}+\boldsymbol{d}$. Then

$$\text{div } (\boldsymbol{D}+\boldsymbol{d}) = \varrho = \text{div } \boldsymbol{D}, \quad \text{div } \boldsymbol{d} = 0; \tag{4.26}$$

also, at surfaces of discontinuity,

$$\boldsymbol{\hat{v}} \cdot [\boldsymbol{D}_+ + \boldsymbol{d}_+ - \boldsymbol{D}_- - \boldsymbol{d}_-] = \sigma = \boldsymbol{\hat{v}} \cdot [\boldsymbol{D}_+ - \boldsymbol{D}_-], \quad \boldsymbol{\hat{v}} \cdot (\boldsymbol{d}_+ - \boldsymbol{d}_-) = 0; \tag{4.27}$$

§ 4.5 DIELECTRICS

and on conducting boundaries

$$\iint (D+d)\cdot \hat{v}\, dS = \iint D\cdot \hat{v}\, dS; \quad \iint d\cdot \hat{v}\, dS = 0. \tag{4.28}$$

$$D+d = \varepsilon(E+e), \quad \text{therefore} \quad d = \varepsilon e. \tag{4.29}$$

From eqns. (4.26) and (4.29) we see that div$(\varepsilon e) = 0$ and so we can write

$$\varepsilon e = \text{curl } \boldsymbol{\theta},$$

where $\boldsymbol{\theta}$ is an arbitrary vector field. This gives the most general form for e which satisfies the conditions (4.26) and (4.29).

The expression for the energy is

$$W+\delta W = \tfrac{1}{2}\iiint \varepsilon(E+e)^2\, d\tau = W + \iiint \varepsilon E\cdot e\, d\tau + \tfrac{1}{2}\iiint \varepsilon e^2\, d\tau.$$

But

$$\iiint \varepsilon E\cdot e\, d\tau = \iiint E\cdot \text{curl } \boldsymbol{\theta}\, d\tau = \iiint [\boldsymbol{\theta}\cdot \text{curl } E + \text{div}(\boldsymbol{\theta}\times E)]\, d\tau.$$

We transform $\iiint \text{div}(\boldsymbol{\theta}\times E)\, d\tau$ into surface integrals over the boundaries or discontinuities,

$$\iiint \text{div}(\boldsymbol{\theta}\times E)\, d\tau = -\sum_L \iint_L \boldsymbol{\theta}\cdot (E_+ - E_-)\times \hat{v}\, dS - \sum_i \iint_{S_i} \boldsymbol{\theta}\cdot (E\times \hat{v})\, dS.$$

Therefore

$$\delta W = \iiint \boldsymbol{\theta}\cdot \text{curl } E\, d\tau - \iint_L \boldsymbol{\theta}\cdot (E_+ - E_-)\times \hat{v}\, dS - \sum_i \iint_{S_i} \boldsymbol{\theta}\cdot (E\times \hat{v})\, dS$$
$$+ \tfrac{1}{2}\iiint \varepsilon e^2\, d\tau.$$

Because $\tfrac{1}{2}\iiint \varepsilon e^2\, d\tau$ must be positive, and because $\boldsymbol{\theta}$ is arbitrary, the necessary and sufficient condition that δW has a minimum is that

$$\text{curl } E = 0,$$
$$(E\times \hat{v}) = 0 \text{ on conducting boundaries,}$$
$$(E_+ - E_-)\times \hat{v} = 0 \text{ on surface charges.}$$

These conditions are those which are satisfied by the actual field in addition to (1)–(4) above. In particular, the condition $E\times \hat{v} = 0$ implies that the potential V, whose existence is shown by $\text{curl } E = 0$, is constant over a conductor. Expressing this "pictorially" we may say that the lines of force "adjust" themselves in the field so that **curl E** = *0*, and the charges arrange themselves on the conductors so that the potential is constant; any other arrangement of field lines and charges on the conductors would give a greater energy.

Example. The introduction of a new insulated uncharged conductor to fill the volume T in an electrostatic field reduces the energy of the field. (The volume T in the original field is free from charge ϱ.)

We take **E'**, **D'** as the final field vectors, **E**, **D** as the initial field vectors.

Outside T, {**E, D**} and {**E', D'**} correspond to the same fixed charges ϱ, σ, Q_l and to the same total charges on conducting surfaces.

On the surface S_T of T, $V' = V'_T$ = constant,

$$\iint_{S_T} \mathbf{D} \cdot \hat{\mathbf{v}} \, dS = 0 = \iint_{S_T} \mathbf{D'} \cdot \hat{\mathbf{v}} \, dS,$$

Inside T, $\quad\quad\quad$ div **D** = 0, $\quad$ **D'** = **E'** = *0*,

$$W' = \tfrac{1}{2} \iiint_{\infty - T} \mathbf{E'} \cdot \mathbf{D'} d\tau, \quad W = \tfrac{1}{2} \iiint_{\infty} \mathbf{E} \cdot \mathbf{D} \, d\tau.$$

Therefore $\quad W' - W = \tfrac{1}{2} \iiint_{\infty - T} (\mathbf{E'} \cdot \mathbf{D'} - \mathbf{E} \cdot \mathbf{D}) \, d\tau - \tfrac{1}{2} \iiint_T \mathbf{E} \cdot \mathbf{D} \, d\tau.$

Since in the volume of the integrations

$$\mathbf{E'} \cdot \mathbf{D'} - \mathbf{E} \cdot \mathbf{D} = \varepsilon(E'^2 - E^2) = (\mathbf{E'} + \mathbf{E}) \cdot (\varepsilon \mathbf{E'} - \varepsilon \mathbf{E}) = (\mathbf{E'} + \mathbf{E}) \cdot (\mathbf{D'} - \mathbf{D}),$$

$$\tfrac{1}{2} \iiint_{\infty-T} (\mathbf{E'}\cdot\mathbf{D'} - \mathbf{E}\cdot\mathbf{D}) \, d\tau = \tfrac{1}{2} \iiint_{\infty-T} (\mathbf{E'}+\mathbf{E})\cdot(\mathbf{D'}-\mathbf{D}) \, d\tau$$

$$= \tfrac{1}{2} \iiint_{\infty-T} (\mathbf{E'}-\mathbf{E})\cdot(\mathbf{D'}-\mathbf{D}) \, d\tau + \iiint_{\infty-T} \mathbf{E'}\cdot(\mathbf{D'}-\mathbf{D}) \, d\tau.$$

Let us consider the integral

$$\iiint_{\infty-T} \mathbf{E'}\cdot(\mathbf{D'}-\mathbf{D}) \, d\tau = -\iiint_{\infty-T} (\mathbf{D'}-\mathbf{D}) \cdot \text{grad } V' \, d\tau$$

$$= \iiint_{\infty-T} [V' \text{ div } (\mathbf{D'}-\mathbf{D}) - \text{div}\{V'(\mathbf{D'}-\mathbf{D})\}] \, d\tau.$$

Everywhere outside T, div $(\mathbf{D'}-\mathbf{D}) = 0$. Therefore

$$\iiint_{\infty-T} \mathbf{E'}\cdot(\mathbf{D'}-\mathbf{D}) \, d\tau = -\iiint_{\infty-T} \text{div } [V'(\mathbf{D'}-\mathbf{D})] \, d\tau$$

$$= \iint_{S_T} V'(\mathbf{D'}-\mathbf{D})\cdot d\mathbf{S} = V'_T \iint_{S_T} (\mathbf{D'}-\mathbf{D}) \, d\mathbf{S} = 0.$$

Therefore $\quad W' - W = -\tfrac{1}{2} \iiint_T \mathbf{E}\cdot\mathbf{D} \, d\tau - \tfrac{1}{2} \iiint_{\infty-T} (\mathbf{E'}-\mathbf{E})\cdot(\mathbf{D'}-\mathbf{D}) \, d\tau < 0$

since both integrands are positive everywhere.

§ 4.6   DIELECTRICS   141

Since the introduction of an uncharged insulated conductor reduces the energy of the system we conclude that such a conductor will be attracted into regions where this reduction is greater, i.e. the attraction will be such as to make the energy least. This means that, in general a conductor is attracted towards parts of the field which are strongest. The smaller the conductor the more exactly is this the case; if the conductor is large, without further investigation of the effect of the size of the conductor on the value of the integrals, we cannot tell what position gives the greatest reduction of energy.

4.6 The electrostatic stress system: Maxwell's stress tensor

The concept of a field, as introduced in Chapter 1, ascribes the effects produced by the field at a point to the state of affairs in the space, or medium, immediately surrounding that point; the charges are the "end-effects" of the field. The original use of this concept, to explain the effects by a mechanical model, led to an elaboration in terms of elasticity which we now describe. But this form of description, although devised for this purpose, is also very valuable in the more modern use of the field described above, for reconciling the theory with modern ideas of space and time. Now the field is pictured by tubes of force which are concentrated according to the strength of the field, and which arise from or end on charges in the field. These charges experience forces, and the field outlook requires us to seek an origin of these forces in the immediate vicinity of the charges: we can find such an origin in the action of the tubes of force ending on the charge provided we assume that each tube tends to contract along its length, as does an elastic string under tension. The resultant of the tensions of all the tubes of force ending on a given charge is the force experienced by that charge.

However, if a tube behaved entirely as an elastic string under tension, it would take up its position in the straight line joining the positive and negative charges on which it begins and ends. The tubes do not do this; some tubes take up longer paths than others. It would appear, therefore, that the tubes tend to push one another apart sideways and their final positions of equilibrium result from their tendencies to contract along their lengths and to repel one another transversely.

Considering any area in the field, we can say that the tubes on one side of this area exert forces across this area on the tubes on the other side. Such an action is similar to the action of one part of an elastic body on another part and is measured by a stress tensor. Before obtaining quantitative expressions for this tensor we examine more closely the application of this idea to the fields pictured.

Figure 2.1(v), p. 27, illustrates the field between two unlike charges and indicates that the tubes close to the charges are more concentrated round that side of the positive charge nearer to the negative charge than on the far side. (Similarly for the tubes ending on the negative charge.) If each tube is assumed to be under tension, then the resultant action of the tubes on either charge is to exert a resultant force on it acting towards the other charge. If we draw a plane area somewhere between the charges so that the charges lie on opposite sides of this plane (e.g. the plane perpendicular to the axis of symmetry of the figure), then the tubes crossing this area on the whole are directed nearly parallel to the normal. Hence the action of the tubes on the left of the plane is to pull those tubes on the right of the plane towards the left; the tension predominates over the sideways repulsion. If we took the resultant of all such actions across all the elements of an infinite plane (or any surface) separating the charges, we should find it to be equal to the attraction between the charges. In other words, the stresses in the tubes transmit the action between the charges through the field much in the way that an elastic body transmits a force from one point to another.

In Fig. 2.1(iv), p. 27, is pictured the field between two like charges. Here the concentration of the tubes on the farther sides of the charges "explains" the repulsion between the charges. Also, across an area separating the charges the tubes are more nearly parallel to the plane, than to the normal. In this case the sideways repulsion between the tubes predominates over the tension; and the force exerted by tubes to the left of the plane on those to the right is a force acting to the right. The resultant of all such actions across all the elements of a surface which separates the charges is equal to the repulsion.

In the above qualitative description we have assumed some values which we obtain below. Once more we warn the reader against reading too much into this picture. The fact that we can give an "explanation" of the forces which are experienced by bodies in an electrostatic field by means of a stress system in the tubes mapping the field does not imply that these tubes have any further properties of real elastic bodies. The above is a picture which gives a method of calculating the forces on bodies in a field, but we have no means of demonstrating the existence of this system apart from verifying the resultant forces on finite bodies. Our picture implies the presence of stresses, even in a vacuum. We now give a quantitative, mathematical representation of the stresses in a field.

The forces which act in any field act not only on charges and conductors, but also on non-uniformities of the permittivity. In order not to obscure the working with too many complications we consider a field in which there is no polarization, i.e. the permittivity is everywhere ε_0, but where there may be a volume distribution of charge in addition to charges on the conductors at the boundary of the field.

§ 4.6 DIELECTRICS

For convenience we use suffix notation and the summation convention (a repeated suffix in any term is assumed to be summed over the values 1, 2, 3; see *C.M.*, vol. 4, §§ 1.11–12). When a small area αn_i, whose positive side is denoted by the unit vector n_i, is subject to a stress T_{ij}, the force exerted by the agent (in this case the tubes of force) situated on the positive side on those situated on the negative side is given by the force f_i where

$$f_i = \alpha T_{ij} n_j. \tag{4.30}$$

This defines the stress T_{ij}. We consider now an arbitrary closed surface G drawn in the field and enclosing only a volume charge ϱ. The only ponderomotive force which acts on charge inside this surface G is given by

$$\mathbf{F} = \iiint_G \sigma \mathbf{E} \, d\tau, \quad \text{i.e.} \quad F_i = \iiint_G \varrho E_i \, d\tau. \tag{4.31}$$

This force must also be given by the stresses in the tubes where they cross G, i.e.

$$F_i = \oiint T_{ij} n_j \, dS.$$

Hence, using the divergence theorem, we deduce

$$\oiint_G T_{ij} n_j \, dS = \iiint_G \frac{\partial T_{ij}}{\partial x_j} \, d\tau = \iiint_G \varrho E_i \, d\tau.$$

In addition to the equality of the resultant forces we must also have that the moments of the two sets of forces about an arbitrary point are equal. (The origin will serve as this arbitrary point.) Therefore

$$\boldsymbol{\Gamma}(O) = \iiint \varrho(\mathbf{r} \times \mathbf{E}) \, d\tau, \quad \text{i.e.} \quad \Gamma_i = \iiint \varepsilon_{ijk} \varrho x_j E_k \, d\tau. \tag{4.32}$$

The moment of the stress system acting at the surface of G is

$$\Gamma_i = \oiint \varepsilon_{ijk} x_j T_{kl} n_l \, dS.$$

Hence, by the divergence theorem,

$$\iiint \frac{\partial}{\partial x_l} (\varepsilon_{ijk} x_j T_{kl}) \, d\tau = \iiint \varepsilon_{ijk} \varrho x_j E_k \, d\tau.$$

Since G is arbitrary we deduce

$$\frac{\partial T_{ij}}{\partial x_j} = \varrho E_i, \quad \frac{\partial}{\partial x_l} (\varepsilon_{ijk} x_j T_{kl}) = \varepsilon_{ijk} \varrho x_j E_k.$$

Since ε_{ijk} is independent of x_i and $\dfrac{\partial x_j}{\partial x_l} = \delta_{jl}$, we deduce that

$$\varepsilon_{ijk}\delta_{jl}T_{kl} + \varepsilon_{ijk}x_j \frac{\partial T_{kl}}{\partial x_l} = \varepsilon_{ijk}T_{kj} + \varepsilon_{ijk}x_j\varrho E_l = \varepsilon_{ijk}\varrho x_j E_k.$$

Therefore
$$\frac{\partial T_{ij}}{\partial x_j} = \varrho E_i, \quad \varepsilon_{ijk}T_{kj} = 0. \tag{4.33, 4}$$

The second of these relations implies that T_{ij} is symmetric in the suffixes. In suffix notation $\varrho = \operatorname{div} \mathbf{D} = \partial D_j/\partial x_j$, so that

$$\frac{\partial T_{ij}}{\partial x_j} = E_i \frac{\partial D_j}{\partial x_j} = \varepsilon_0 \frac{\partial}{\partial x_j}(E_i E_j) - \varepsilon_0 E_j \frac{\partial E_i}{\partial x_j}$$

$$= \varepsilon_0 \frac{\partial}{\partial x_j}(E_i E_j) - \varepsilon_0 E_j \left(\frac{\partial E_i}{\partial x_i} - \frac{\partial E_j}{\partial x_i}\right) - \varepsilon_0 E_j \frac{\partial E_j}{\partial x_i}.$$

But since **curl** $\mathbf{E} = \mathbf{0}$, the bracket in the middle term is zero. Therefore

$$\frac{\partial T_{ij}}{\partial x_j} = \varepsilon_0 \frac{\partial}{\partial x_j}(E_i E_j) - \varepsilon_0 \frac{\partial}{\partial x_i}\left(\frac{1}{2} E_k E_k\right).$$

We can write this relation thus

$$\frac{\partial T_{ij}}{\partial x_j} = \frac{\partial}{\partial x_j}\left(\varepsilon_0 E_i E_j - \frac{1}{2}\varepsilon_0 E_k E_k \delta_{ij}\right).$$

We deduce, as a possible form for the stress tensor,

$$T_{ij} = \varepsilon_0 (E_i E_j - \tfrac{1}{2} \delta_{ij} E_k E_k). \tag{4.35}$$

Of course, this choice of T_{ij} is not unique; we could add to it any function, including a constant, whose divergence vanishes. However, (4.35) is the form adopted to give *Maxwell's stress tensor*. Written explicitly,

$$T_{ij} = \begin{pmatrix} \tfrac{1}{2}\varepsilon_0(E_1^2 - E_2^2 - E_3^2) & \varepsilon_0 E_1 E_2 & \varepsilon_0 E_1 E_3 \\ \varepsilon_0 E_2 E_1 & \tfrac{1}{2}\varepsilon_0(-E_1^2 + E_2^2 - E_3^2) & \varepsilon_0 E_2 E_3 \\ \varepsilon_0 E_3 E_1 & \varepsilon_0 E_3 E_2 & \tfrac{1}{2}\varepsilon_0(-E_1^2 - E_2^2 + E_3^2) \end{pmatrix}. \tag{4.36}$$

If now we choose the x_1-axis to coincide with the direction of the field, so that $E_i = \{E\ 0\ 0\}$, and use an area with $n_i = \{1\ 0\ 0\}$,

$$T_{ij} = \begin{pmatrix} \tfrac{1}{2}\varepsilon_0 E^2 & 0 & 0 \\ 0 & -\tfrac{1}{2}\varepsilon_0 E^2 & 0 \\ 0 & 0 & -\tfrac{1}{2}\varepsilon_0 E^2 \end{pmatrix}. \tag{4.37}$$

§ 4.6   DIELECTRICS   145

This implies that the force acting across an area at right-angles to the lines of force is parallel to the field, i.e.
$$f_i = \alpha T_{ij}n_j = \alpha T_{i1} = \alpha(\tfrac{1}{2}\varepsilon_0 E^2,\ 0,\ 0).$$

When the coordinate axes have been rotated to coincide with the direction of E and the orthogonal directions, the stress tensor is reduced to the diagonal form (4.37). These diagonal terms are the *principal stresses* and the coordinate directions at that point are the *principal* directions. We can also regard the stress system, given in eqn. (4.37) as a uniform hydrostatic pressure of magnitude $\tfrac{1}{2}\varepsilon_0 E^2$, corresponding to $-\tfrac{1}{2}\varepsilon_0 E^2$ along the diagonal, together with a tension $\varepsilon_0 E^2$ along the direction of E, an additional term $+\varepsilon_0 E^2$ in the leading position in eqn. (4.36). As a special case of this result we consider the field close to a conducting boundary, where $D = \sigma\hat{n} = \varepsilon_0 E$. Then
$$E_i = \{\sigma/\varepsilon_0\ 0\ 0\},\quad f_i = \alpha\{(\sigma^2/2\varepsilon_0)\ 0\ 0\},$$
which is the usual expression for the force on a conductor.

When the area is parallel to the field, e.g. $n_i = \{0\ 1\ 0\}$,
$$f_i = \alpha T_{ij}n_j = \alpha T_{i2} = \alpha\{0\ -\tfrac{1}{2}\varepsilon_0 E^2\ 0\}.$$

This shows the transverse repulsion between the tubes of force.

We give now a vector form for the force f transmitted across a small area $\alpha\hat{n}$. From eqns. (4.30) and (4.35)
$$f_i = \alpha T_{ij}x_j = \alpha\varepsilon_0(E_iE_j - \tfrac{1}{2}\delta_{ij}E_kE_k)n_j = \alpha\varepsilon_0(E_iE_jn_j - \tfrac{1}{2}E_kE_kn_i).$$
Therefore $f = \alpha\varepsilon_0\{E(E\cdot\hat{n}) - \tfrac{1}{2}E^2\hat{n}\} = \alpha\varepsilon_0\{E\times(E\times\hat{n}) + \tfrac{1}{2}E^2\hat{n}\}.$

Example. We calculate the repulsion between two equal charges Q, Fig. 4.6.

At P, using cylindrical polars, the field of the charges is
$$E = \{E_r,\ E_\theta,\ E_z\} = \frac{1}{4\pi\varepsilon_0}\left\{\frac{2Qr}{(r^2+a^2)^{3/2}}\ 0\ 0\right\}.$$

Hence across the element of area $r\,dr\,d\theta$ near P the force exerted by the tubes in the region

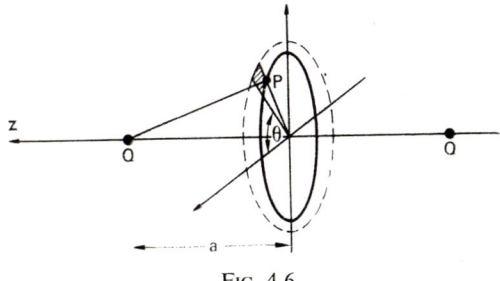

FIG. 4.6

$z > 0$ is

$$\mathbf{dF} = \frac{r\,dr\,d\theta}{8\pi^2\varepsilon_0}\left\{0\quad 0\quad \frac{Q^2 r^2}{(r^2+a^2)^3}\right\} = \{0\quad 0\quad dZ\}.$$

Therefore
$$Z = 2\pi \int_0^\infty \frac{Q^2}{8\pi^2\varepsilon_0}\cdot\frac{r^3\,dr}{(r^2+a^2)^3} = \frac{Q^2}{4\pi\varepsilon_0}\cdot\frac{1}{(2a)^2}.$$

after using the substitution $r = a\tan\theta$ to perform the integration. This is the force of repulsion between the charges.

The concept of tubes of force under tension also fits in with the results concerning the energy of the electrostatic field. This energy is the energy "stored" in the stretched tubes of force. We have seen (cf. the example of p. 140) that the insertion of an uncharged insulated conductor reduces the energy of the field; this energy reduction is due to shortening of the tubes. Again, the presence of a dielectric reduces the energy, and it does so by "reducing the tension" in the tubes of force; for example, in a parallel plate condenser, the displacement $\mathbf{D}$ is unaltered by inserting dielectric but the field $\mathbf{E}$ (and so the potential difference) is reduced.

Finally, the parallel between the expressions $\frac{1}{2}xT$ for the energy in a string at tension T and extension (displacement) x and the expression $\frac{1}{2}\mathbf{D}\cdot\mathbf{E}$ for the energy density of a field of strength $\mathbf{E}$ (cf. tension) and displacement $\mathbf{D}$ (cf. extension) supports the choice of $\frac{1}{2}\mathbf{D}\cdot\mathbf{E}$ as the energy density. We point out once again that both the expressions for the energy density and the stress are hypotheses which are adopted because of their simple forms. These expressions are not unique, but on integration they give results closely approximating to those found in practice.

Miscellaneous Exercises IV

1. The inner conductor of a coaxial cable is a long straight wire whose cross-section is a circle of radius a. The wire is covered with an insulating sheath of uniform thickness d and dielectric constant K. Outside this sheath is a hollow thin cylindrical conductor of radius c ($c > a+d$). Find the capacitance of the cable per unit length.
2. Obtain the values of the potential and electric field at points inside and outside a sphere of radius a and dielectric constant K which is uniformly charged with density ϱ.
3. (i) Three closed surfaces 1, 2, 3 are equipotentials in an electric field. If the space between 1 and 2 is filled with a dielectric K, and that between 2 and 3 is filled with a dielectric K', show that the capacity of a condenser having 1 and 3 for faces is C, given by

$$\frac{1}{C} = \frac{1}{AK} + \frac{1}{BK'},$$

where A, B are the capacities of air condensers having as faces the surfaces 1, 2 and 2, 3 respectively.

§ 4.6 DIELECTRICS

(ii) A number of surfaces $S_1, S_2, \ldots, S_n$ are closed equipotentials in a certain electric field and are such that S_p lies entirely within S_q if $p < q$. The capacity of a condenser formed by two thin conductors coinciding with S_1 and S_2 is denoted by C_{12}, and so on. If a condenser is formed by two conductors coinciding with S_1 and S_n, and the space between S_1 and S_2 is filled with uniform dielectric K_{12}, that between S_2 and S_3 by dielectric K_{23}, and so on, find the resulting capacity.

4. A heterogeneous dielectric is formed of n concentric spherical layers of specific inductive capacities $K_1, K_2, \ldots, K_n$, starting from the innermost dielectric, which forms a solid sphere; also the outermost dielectric extends to infinity. The radii of the spherical boundary surfaces are $a_1, a_2, \ldots, a_{n-1}$ respectively. Prove that the potential due to a quantity Q of electricity at the centre of the spheres at a point distant r from the centre in the dielectric K_s is

$$\frac{1}{4\pi\varepsilon_0}\left\{\frac{Q}{K_s}\left(\frac{1}{r}-\frac{1}{a_s}\right)+\frac{Q}{K_{s+1}}\left(\frac{1}{a_s}-\frac{1}{a_s+1}\right)+\ldots+\frac{Q}{K_n}\frac{1}{a_n}\right\}.$$

5. A condenser is formed by two concentric conducting spherical surfaces of radii a and b ($a < b$). The space between is filled with a medium whose dielectric constant at any point P is $K_0\varrho/(b-a)$, where K_0 is constant and ϱ is the distance of P from a certain fixed diameter of the spheres. Show that the lines of force are radial and that the capacity of the condenser is

$$2\pi^2\varepsilon_0 K_0 a^2 b^2/(b-a)^2(b+a).$$

6. Write down the conditions satisfied by the displacement D and the electric intensity E in the field due to n closed conductors with surfaces $S_1, S_2, \ldots, S_n$ and charges $Q_1, Q_2, \ldots, Q_n$ in an unbounded uncharged medium in which the dielectric coefficient K is a function of position. Show that the energy of the field is

$$\frac{1}{2}\iiint_V E\cdot D\,d\tau,$$

where V denotes the volume outside the conductors.
Vectors D', E' are such that

$$\text{div } D' = 0, \quad D' = KE'$$

at all points of V and such that

$$\oiint_{(S_k)} D'\cdot dS_k = Q_k \quad (k = 1, 2, \ldots, n),$$

where dS_k is in the direction of the normal to S_k drawn into the medium; prove that

$$\iiint_V E'\cdot D'\,d\tau \geqslant \iiint_V E\cdot D\,d\tau$$

if D', E' are of order $1/r^2$ at a great distance r from the conductors.

7. Show that the electrostatic energy of a system of n charged conductors is $\frac{1}{2}\sum_{i=1}^{n} Q_i V_i$, Q_i being the charge and V_i the potential of the ith conductor. Prove that this energy is equivalent to $\frac{\varepsilon_0}{2}\int KE^2\,d\tau$ taken over all space, E being the electric intensity at any point of the field.
If the charges on the conductors are redistributed in any manner so that the total charge

on each conductor remains the same, but the conductors do not necessarily remain equipotentials, and E' is the electric intensity arising from this new distribution, show that

$$\int K\mathbf{E}\cdot\mathbf{E}'\,d\tau = \int KE^2\,d\tau.$$

Deduce that
$$\int KE^2\,d\tau \leq \int KE'^2\,d\tau.$$

The total charge on each conductor being given, show that the energy of the field is least if the distribution of charge on all surfaces is such that they are equipotentials.

8. By applying Green's theorem to div $(K\delta V \operatorname{\mathbf{grad}} V)$ prove that the energy of the electrostatic field due to a number of charged conductors in the presence of isotropic dielectric material is decreased when there is a small increase in the dielectric constant K over any region, provided the total charge on each conductor remains unaltered.

Show that the energy of the field increases when there is a similar change in the dielectric constant and the potentials of the conductors are kept constant.

CHAPTER 5

THE STEADY FLOW OF ELECTRIC CURRENTS

5.1 Introduction

In the electrostatic fields already considered charges were stationary and a conductor, because charge can move inside it, was a region inside which $E = 0$, i.e. $V =$ constant. Now we relax this restriction of stationary charges and consider steady motion of charges.

When water in a lake, or tank, is motionless the surface is horizontal but when water is running *steadily* over a weir or down a stream, the surface remains stationary but is not necessarily horizontal and the velocity of the water at any point does not vary with time. (This is the implication of the term *steady motion*.) This picture illustrates the state of affairs in electricity. For the static case the potential of the conductor is the same at all points —corresponding to the horizontal surface of the lake. When two points of a conductor are maintained at different, but constant, potentials a steady electric current flows in the conductor and each point takes up its own potential— corresponding to the irregular surface of the water in a stream.

In order to maintain a steady flow down a stream, water must be supplied continually at the top, or other points, and be removed continually lower down; or water must be made to circulate continually by the use of a pump which raises the water back to the higher level. In the electrical case, either current enters a conductor at one or more points and leaves it at others— such points or regions are *electrodes*—or it flows around a closed conducting circuit under the influence of a battery, which corresponds to the pump. We shall refer later to this picture of the flow of water in order to illustrate the behaviour of electric currents.

We defined charge as a basic unit in Chapter 1 so we now define the current strength I in a conducting wire as the rate of passage of charge past any point

of the wire; in symbols
$$I = dQ/dt, \qquad (5.1)$$

where I is measured in amperes and Q in coulombs. (In fact, because it is easier to reproduce a steady current exactly than it is to reproduce an exact charge, the international system of units takes current in preference to charge as its basic unit. Therefore eqn. (5.1) is used to define Q from I, by integration. This is the origin of the former nomenclature MKSA for this system of units, A standing for the ampere, which is the fourth basic unit.)

Now we define the *current density vector* $\boldsymbol{j}$. When, in the course of the motion, charge crosses an arbitrary (small) area $\alpha \hat{e}$ from the negative to the positive side (as defined by $\hat{e}$) at the rate dQ/dt, we define $\boldsymbol{j}$ by the relation

$$\boldsymbol{j} \cdot \hat{e} = \lim_{\alpha \to 0} \frac{1}{\alpha} \frac{dQ}{dt}. \qquad (5.2)$$

[For reasons which will be clear later the reader should compare (5.2) with the definition of $\boldsymbol{D}$ in eqn. (2.9).] In this way we define the vector $\boldsymbol{j}$ at every point of a conductor and can represent the flow pattern by *tubes of flow*. These are surfaces generated by stream-lines, i.e. field lines of $\boldsymbol{j}$, drawn through the perimeters of arbitrary areas. We shall refer to *current filaments* as tubes of flow with a small cross-sectional area α (which may vary from point to point along the filament), the strength of the filament being given by the current flowing along it, viz.

$$I = |\boldsymbol{j}|\alpha.$$

A common example of a current filament is a wire carrying a current I. However, we shall regard a region of current flow as filled with tubes of flow much in the way in which we regarded an electrostatic field as "filled" with tubes of force.

First we consider a region in which current enters and leaves across the boundary of the region and consider later the generation of currents in a region.

5.2 The equation of continuity (conservation) of charge

Positive and negative charges can neutralize one another but the total electric charge, taking account of sign, in a closed system cannot alter. This is the principle of conservation of charge, which we assume to hold universally. If the total charge inside a closed surface does alter, it can only do so because charge enters the region across this surface. We express this analytically by considering the volume T inside an arbitrary closed surface G drawn

§ 5.3   THE STEADY FLOW OF ELECTRIC CURRENTS

inside a conductor, the conductor being free from sources of charge. If the charge density at any point is ϱ, then the total charge inside G is

$$Q = \iiint_T \varrho \, d\tau.$$

Keeping G fixed, we deduce that

$$\frac{dQ}{dt} = \iiint_T \frac{\partial \varrho}{\partial t} \, d\tau.$$

But, from eqn. (5.2) we can also write

$$\frac{dQ}{dt} = -\oiint_G \boldsymbol{j} \cdot \boldsymbol{dS},$$

where $\boldsymbol{j}$ is the current vector and $\boldsymbol{dS}$ is directed out of T. Therefore

$$\iiint_T \frac{\partial \varrho}{\partial t} \, d\tau + \oiint_G \boldsymbol{j} \cdot \boldsymbol{dS} = \iiint_T \left\{ \frac{\partial \varrho}{\partial t} + \operatorname{div} \boldsymbol{j} \right\} d\tau = 0.$$

This integral must vanish for an arbitrary surface G, so that

$$\frac{\partial \varrho}{\partial t} + \operatorname{div} \boldsymbol{j} = 0. \tag{5.3}$$

This *equation of continuity* must hold at all points of a conductor free from sources of current. Because we consider only *steady* conditions $\partial \varrho / \partial t = 0$. Therefore

$$\operatorname{div} \boldsymbol{j} = 0. \tag{5.4}$$

5.3 Ohm's law

The well-known elementary form of Ohm's law states that when a conductor carries a current I from a point A, at potential V_A, to a point B, at potential V_B, then

$$V_A - V_B = RI, \tag{5.5}$$

where R is the *resistance* of the conductor between A and B and depends only on the shape, material (and temperature) of the conductor. When V_A, V_B are measured in volts and I in amperes then R is measured in *ohms* (symbol Ω).

When current flows inside a conducting region the flow can be divided into filaments of small cross-section and, because $\operatorname{div} \boldsymbol{j} = 0$, these filaments must begin and end on electrodes (cf. the electrostatic field in a charge-free region, in which div $\mathbf{D} = 0$, bounded by conductors). We consider points A, B separated by a small displacement $\hat{e}s$ along a filament of small cross-section α. Then the current flowing between them is $I = |\mathbf{j}|\alpha$ and $\mathbf{j}$ has the same direction as $\hat{e}$, everywhere along the filament. Hence, Ohm's law applied to this small element gives

$$IR = R|\mathbf{j}|\alpha = V_A - V_B = -s\hat{e} \cdot \mathbf{grad}\, V.$$

Since the different points of the conductor are at different potentials there is an electrostatic field inside the conductor given by $\mathbf{E} = -\mathbf{grad}\, V$. Therefore,

$$|\mathbf{j}| = s/(R\alpha)\hat{e} \cdot \mathbf{E}, \quad \text{i.e.}\ \mathbf{j} = \sigma\mathbf{E}, \tag{5.6}$$

since $\mathbf{j}$ and $\hat{e}$ are everywhere in the same direction. The equation (5.6) is the local, or field, form of *Ohm's law*, the quantity σ, the *conductivity*, depending on the material of the conductor (cf. the equation $\mathbf{D} = \varepsilon\mathbf{E}$ for an electrostatic field).

Example. We calculate the equivalent resistance of a set of resistors in series.

The relation between R and σ in eqn. (5.6) shows that, if A and B are at a finite distance apart along the filament, then

$$R_{AB} = \int_A^B \frac{ds}{\sigma\alpha}$$

is the resistance of the filament joining A, B, the integral being a line integral taken along the filament. If $P_1, P_2, \ldots, P_{n-1}$ are a number of points lying on the filament between A, B, then

$$R_{AB} = \int_A^{P_1} + \int_{P_1}^{P_2} + \ldots + \int_{P_{n-1}}^B \frac{ds}{\sigma\alpha}$$
$$= R_1 + R_2 + \ldots + R_n,$$

$R_1, R_2, \ldots, R_n$ being the resistances of the various segments of AB which are connected in series. [See also example (2), p. 174.]

Equation (5.6) leads to the differential equation for V, since div $\mathbf{j} = 0$,

$$\text{div}\,(\sigma\,\mathbf{grad}\,V) = 0, \tag{5.7}$$

or, if σ is uniform,

$$\nabla^2 V = 0. \tag{5.8}$$

Equations (5.4), (5.6) and (5.7) or (5.8) are the field equations which govern the steady flow of electric currents.

According to the ideas of the structure of matter outlined in Chapter 1, a conductor, either metallic or electrolytic, contains charged particles, electrons or ions, in motion, and these particles have different charges and different velocities. Suppose that there are n particles in unit volume each carrying a

§ 5.4 THE STEADY FLOW OF ELECTRIC CURRENTS

charge q and each having a velocity v. (The density n is, in general, a function—called a distribution function in probability theory—of position, q and v.) Those particles of this group which pass through α in a short time δt lie inside a volume $\alpha(\hat{e} \cdot v)\delta t$ (see Fig. 5.1). Hence their number is $n\alpha(\hat{e} \cdot \hat{v})\delta t$ and the charge carried across α by them is $nq\alpha(\hat{e} \cdot v)\delta t$. Hence, from the definition of j,

$$\hat{e} \cdot j = \sum nq(\hat{e} \cdot v),$$

where $\sum$ denotes the sum, or integration, over all possible values of q and v. Since $\hat{e}$ is arbitrary we deduce that

$$j = \sum nqv. \tag{5.9}$$

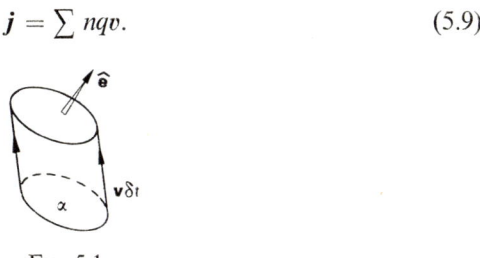

Fig. 5.1

When there is a potential difference between different points of a conductor, these charged particles are moving in an electric field and so experience a force. This force accelerates a particle until it collides with another particle, perhaps a stationary atom of the metallic framework, or another moving particle or ion. The average effect of these accelerations and collisions is that the charged particles drift in the field with a mean velocity in the direction of the electric field—or exactly opposite if the charge is negative. In a region of the conductor where there is no electric field, $j = 0$, because the particle velocities are randomly oriented. But when there is an electric field the sum $\sum qnv$ has a resultant in the direction of E because of the average velocity of drift. Hence, our rough ideas suggest the relation (5.6), Ohm's law.

5.4 Boundary conditions and discontinuities

So far we have derived the equations which govern the flow of a current inside the conducting region; now we obtain those which apply at boundaries or other surfaces of discontinuity. We assume, of course, that the conditions apertaining to the electrostatic field are the same as those obtained in earlier chapters.

An electrode is the point or region of the surface of a conductor where current enters or leaves. In practice such electrodes are usually made of material of high conductivity such as copper, so we regard an electrode as the limiting case of the practical arrangement. At an electrode, then, current enters the given region from another conductor which has infinite conductivity, the current flowing across the boundary surface. Since the current strength is finite and $j = \sigma E$ we must take $E = 0$, or $V =$ constant in the material of the electrode. In the given conductor therefore the current must enter, or leave, in a direction normal to the surface of the electrode, and all points of this surface have the same potential. In symbols, we have at an electrode

$$\hat{\nu} \times j = 0, \quad V = \text{constant}. \tag{5.10}$$

The *strength* of the electrode is given by the rate at which charge leaves the electrode and enters the conductor, i.e. the strength is measured (in amperes) by

$$I = -\iint j \cdot \hat{\nu} \, dS = \iint \sigma \frac{\partial V}{\partial \nu} \, dS. \tag{5.11}$$

Here and in eqn. (5.10) $\hat{\nu}$ is in the direction of the normal drawn out of the material of the conductor, and the integration is taken over the surface of the electrode (cf. the charge carried on a conductor bounding an electrostatic field).

A conductor need not be a solid but, except at electrodes, it is bounded by non-conducting regions and at all such boundary points

$$j \cdot \hat{\nu} = 0, \quad \frac{\partial V}{\partial \nu} = 0. \tag{5.12}$$

Finally we consider the situation where there is a discontinuity in σ occurring at a surface which separates two regions of different conductivities. The conservation of charge requires that charge is carried away from one side at the same rate as it is carried up to the other side, i.e.

$$\hat{\nu} \cdot (j_+ - j_-) = 0. \tag{5.13}$$

This equation, together with the first of eqn. (5.12), is the surface form of the equation $\text{div } j = 0$, eqn. (5.4). Since Ohm's law is satisfied on either side of the surface, eqn. (5.13) leads to the condition

$$\left(\sigma \frac{\partial V}{\partial \nu}\right)_+ - \left(\sigma \frac{\partial V}{\partial \nu}\right)_- = 0. \tag{5.14}$$

In eqns. (5.13) and (5.14) the direction of the unit normal $\hat{\nu}$ is from the negative side towards the positive side of the surface of discontinuity.

In addition to these conditions the usual conditions for an electrostatic field must be satisfied at the discontinuity. Because a conducting medium cannot (usually) be polarized—the theory of polarization requires that the charges constituting the material are bound elastically to equilibrium positions and so are unable to move as they must when currents flow—we take $D = \varepsilon_0 E$ in conducting media. Then, at a discontinuity there is a stationary distribution of surface charge given by

$$\hat{v} \cdot (D_+ - D_-) = \varepsilon_0 \hat{v} \cdot (E_+ - E_-) = \varepsilon_0 \hat{v} \cdot [(j/\sigma)_+ - (j/\sigma)_-] \neq 0.$$

The remaining condition satisfied by the electrostatic field is

$$\hat{v} \times (E_+ - E_-) = 0 \quad \text{or} \quad V_+ - V_- = \text{constant}. \tag{5.15}$$

5.5 The rate of heat production

It is well known that when a resistance carries a current heat is produced in the resistor. The rate of production of heat, from the definition of charge and potential, is $(V_A - V_B)I$ or $I^2 R$ when a current I flows between A and B. If we consider, instead of a wire, a portion of a current filament of (short) length s and cross-section α, then

$$I = |j|\alpha, \quad V_A - V_B = -s\frac{\partial V}{\partial s} = |E|s = \frac{|j|}{\sigma}s,$$

and the rate of heat production is

$$(V_A - V_B)I = |j|^2 \alpha s/\sigma = (j^2/\sigma)\tau, \tag{5.16}$$

where $\tau \,(= \alpha s)$ is the volume of the (small) conducting element. This indicates a rate of production of heat j^2/σ per unit volume.

It is easy to account for the production of heat by the passage of a current in terms of our rough picture of moving charged particles. The resistance to the passage of the current arises from collisions between the various particles; when a particle, having been accelerated by the electric field, and so having acquired some kinetic energy, collides with another the kinetic energy is shared between the two. These collisions must cause a general, random increase in the velocities of the particles of the material and this increase is apparent as a rise of temperature of the material as a whole. The result that $(j^2/\sigma)\tau$ is the rate of heat production in an arbitrary small volume element τ implies that this balance between loss of electrical potential energy and production of heat is maintained locally at every point.

5.6 Electromotive force

In the water flow analogy given on p. 149 we pointed out that a pump is needed to make water circulate; the water then flows in one or more closed circuits and in order to maintain the steady flow the agent working the pump supplies energy continually to the system.

In electric current flow the place of the pump is taken by some agent to which the general name *electromotive force* (or field) is given. In practice such an agent is a battery or cell which provides energy to the system at the expense of its own chemical energy, a dynamo which converts mechanical energy, or an arrangement such as a thermo-couple, etc. We discuss the action of a dynamo later when we consider electromagnetic induction. Here we give a brief account of two kinds of galvanic, or electrolytic, cell.

When a metal is dipped into a solution of one of its own salts, chemical action (which we do not discuss here) tends to make atoms of the metal enter the solution as ions carrying a charge. When such an atom enters the solution taking say, a positive charge, it leaves an equal negative charge on the plate. The positive ions so produced tend to form a layer of positive charges in the solution and are held near the surface of the plate by the attraction of the residual negative charges. The process of solution continues until the chemical forces tending to produce solution are balanced by the electrical attraction between the layers of charge. If the ions are unable to leave the plate, the equilibrium state corresponds to a double layer of charge at the surface of separation. As we have seen (§ 2.12) such a double layer corresponds to a discontinuity in the potential. If plates of two different metals are dipped into the same electrolyte and connected together by a wire outside the solution, then negative charges can leave plate A, travel along the wire to plate B and there attract an ion of B out of solution, provided that the chemical forces tending to dissolve A are greater than those tending to dissolve B. In this way a galvanic cell produces a current in the external circuit accompanied by the deposition of a substance on one electrode and solution of the other.

Another arrangement consists of two electrodes of the same material dipping into a solution whose concentration is not uniform. Leaving aside considerations of the double layer produced as described above the process of diffusion tends to make ions travel towards regions of lower concentration: this process makes ions of both signs travel in the *same* direction whereas electrical forces make such ions travel in opposite directions. The result of the two processes is that a current flows in an external connection between the plates as long as the concentration is not uniform.

§ 5.6   THE STEADY FLOW OF ELECTRIC CURRENTS   157

Because we appeal to our ideas of the physical microscopic structure of the solution the above discussions of electromotive force are hypothetical. However, for the purposes of our macroscopic theory we regard an electromotive agent as one producing *either* a discontinuity in the potential at some points of a circuit, *or* producing an apparent force, in addition to electrical forces acting on the charges moving in the circuit. These forces supply the energy to the electrical circuit by moving the charges across the discontinuity of the potential, or in some other way moving them against the electrical forces.

We *define* the strength of the electromotive force along a curve between two points A, B in a conducting region to be ψ_{AB} in the equation

$$W_{AB} = \psi_{AB} I \tag{5.17}$$

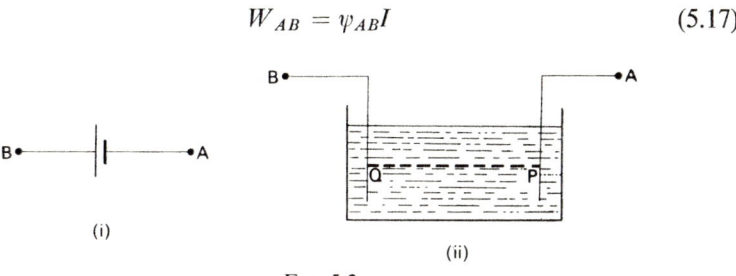

Fig. 5.2

where W_{AB} is the rate of supply of energy to the circuit by the agent when a current I flows along the given curve from A to B. We consider the two cases discussed above. In the first case [see Fig. 5.2(i)] the chemical forces have to provide the energy to carry the charges constituting the current I from the low to the high potential side of the double layer; this requires energy to be expended at the rate $I(V_+ - V_-) = \psi I$. Hence the *electromotive force* (e.m.f.) of the double layer is the same as the discontinuity in the potential. In the second case a charge q, whilst in the electrolyte under the influence of the electromotive field experiences a force $q\mathbf{E}$ from the electric field which must be present when two plates are at different potentials, and a force $q\mathcal{E}$ from the "agent". When the charges follow a path such as $APQB$ in Fig. 5.2(ii) energy is supplied by the "agent" at a rate

$$W_{AB} = \psi_{AB} I = \int_A^B I\mathcal{E} \cdot \mathbf{ds}.$$

Therefore

$$\psi_{AB} = \int_A^B \mathcal{E} \cdot \mathbf{ds}. \tag{5.18}$$

Since $\mathcal{E}$ is zero for this path everywhere except between P and Q, we can write (neglecting any double layers at the electrode)

$$\psi_{AB} = \int_A^B \mathcal{E} \cdot d\mathbf{s} = \int_P^Q \mathcal{E} \cdot d\mathbf{s}.$$

The fundamental significance of this result is that, in general, an electromotive force ψ_{AB} depends upon the path used in the integral, i.e. an electromotive force must be associated with a circuit, or part of one.

To take account of electromotive fields we generalize the field form of Ohm's law, eqn. (5.6), to include their effects and write

$$\mathbf{j} = \sigma(\mathbf{E} + \mathcal{E}). \qquad (5.19)$$

We must remember that $\mathbf{E}\, (= -\mathbf{grad}\, V)$ is an electrostatic field, but that $\mathcal{E}$ is different from zero only in certain regions where some "agent" is at work.

5.7 The analogy with the electrostatic field

There is a close correspondence between the properties (and equations) of an electrostatic field and those of a current distribution. Briefly the correspondence is this: an electrostatic field in a region containing a dielectric produces a displacement $\mathbf{D}$; an electrostatic field inside a conducting region produces a steady flow of current $\mathbf{j}$. The correspondence is shown below.

Current flow	Electrostatics
Electrode	Conductor
Current strength I	Charge Q
Current density $\mathbf{j}$	Displacement $\mathbf{D}$
Conductivity σ	Permittivity ε
$\mathbf{j} = \sigma \mathbf{E}$	$\mathbf{D} = \varepsilon \mathbf{E}$
Conservation of charge div $\mathbf{j} = 0$	Inverse square law div $\mathbf{D} = \varrho$.

The last line above shows that the complete analogy should be drawn with an electrostatic field in a charge-free region ($\varrho = 0$) only. We can pursue the analogy further if we allow the existence of *permanent* polarization of the medium in electrostatics, in addition to the induced polarization included in $\mathbf{D}$.

e.m.f., $\psi_{AB} = V_+ - V_-$	double layer $\tilde{\omega} = V_+ - V_-$
electromotive field $\mathcal{E}$	permanent polarization $(\mathbf{P}/\varepsilon)$
$\mathbf{j} = \sigma(\mathbf{E} + \mathcal{E})$	$\mathbf{D} = \varepsilon \mathbf{E} + \mathbf{P}.$

It is inadvisable to carry this analogy too far; nevertheless the theorems and methods of electrostatics can be used as a guide in the similar context of current flow.

The commonest problem in connection with current flow is the determination of the equivalent resistance of a conducting medium situated between two electrodes. This problem usually reduces to finding a potential function V which satisfies the field equations and boundary conditions and corresponds to potentials V_A and V_B for the electrodes and current strengths I and $-I$ respectively. Then the equivalent resistance is $R = (V_A - V_B)/I$. (There are few problems concerning sources of e.m.f. in the body of a conductor.) Thus such problems are the exact analogues of finding the capacitance of an arrangement of conductors in an electrostatic field. [We can add two more pairs to our list of analogues

Equivalent resistance, $R = (V_A - V_B)/I$ Capacitance, $(1/C) = (V_A - V_B)/Q$

Rate of heat production, VI, V^2/R, I^2R Energy of charged system, $\frac{1}{2}VQ$, $\frac{1}{2}CV^2$, $\frac{1}{2}Q^2/C$

the reciprocal of capacitance being analogous to resistance.]

Experience in solving electrostatic problems helps in the solution of the corresponding current flow problems (and vice versa sometimes,). This is illustrated in the following worked examples.

Example 1. Two concentric spheres $r = a$ and $r = b$ ($a > b$) are separated by material whose specific conductivity σ is $f(\theta, \phi)$. Show that the resistance between the spheres is

$$\frac{a-b}{ab} \bigg/ \left\{ \int_0^{2\pi} d\phi \int_0^{\pi} f(\theta, \phi) \sin\theta \, d\theta \right\}.$$

Since the spheres $r = a$, $r = b$ are equipotentials and the conductivity is independent of r, we try a solution of the equation div $(\sigma \text{ grad } V) = 0$ of the form $V = V(r)$. Then, cf. example 2, p. 123,

$$\frac{d}{dr}\left(r^2 \frac{dV}{dr}\right) = 0, \quad V = A + \frac{B}{r}$$

and the boundary conditions $(V)_{r=a} = V_A$, $(V)_{r=b} = V_B$ give

$$V = \frac{aV_A - bV_B}{a-b} + \frac{(V_B - V_A)ab}{(a-b)r}.$$

[We take $V_A < V_B$ without loss of generality. Note also that the current is purely radial.] The current density vector at $r = b$ is

$$\mathbf{j} = \{j_r \ 0 \ 0\} = -\left\{\sigma \frac{\partial V}{\partial r} \ 0 \ 0\right\} = \left\{\frac{\sigma(V_B - V_A)a}{(a-b)b} \ 0 \ 0\right\}.$$

The current leaving the sphere $r = b$ is

$$I = \int_{r=b} j_r \, dS = b^2 \int_0^{2\pi} d\phi \int_0^{\pi} j_r \sin\theta \, d\theta$$
$$= \frac{(V_B - V_A)ab}{(a-b)} \int_0^{2\pi} d\phi \int_0^{\pi} f(\theta, \phi) \sin\theta \, d\theta.$$

[This is also the current entering the sphere $r = a$.] The resistance between the spheres, R, is defined by $R = (V_B - V_A)/I$ and the required result follows.

Note that, if σ is constant, the resistance is $(a-b)/(4\pi ab\sigma)$, which follows directly from the result concerning equivalent resistance and capacitance of p. 159 since the capacity of a condenser made up of the two concentric spherical conductors *in vacuo* is $4\pi\varepsilon_0 ab/(a-b)$ (see p. 86).

Example 2. Electric current flows through a conductor consisting of a sphere of copper (specific conductivity σ) embedded in an infinite mass of iron (specific conductivity σ_0). The flow at an infinite distance from the sphere is uniform, the current vector being j_0. Prove that the flow through the sphere is uniform and that the current vector there is

$$3\sigma(2\sigma_0 + \sigma)^{-1} j_0.$$

With the usual notation for spherical polar coordinates, we take O at the centre of the sphere and Oz along the direction of the undisturbed current vector j_0 so that, referred to rectangular cartesian axes $Oxyz$, $j_0 = \{0 \ 0 \ j_0\}$. The electric field E_0 at great distances is determined by $j_0 = \sigma_0 E_0$, see Fig. 5.3. Therefore we can take the undisturbed electric potential as $V_0 = -E_0 z = -E_0 r \cos\theta$. This suggests that we take as the total potential

$$V = Ar\cos\theta \quad \text{for} \quad r \leqslant a,$$
$$V = -E_0 r \cos\theta + \frac{B\cos\theta}{r^2} \quad \text{for} \quad r \geqslant a,$$

where A, B are constants to be determined. These potentials satisfy Laplace's equation, introduce no singularities and satisfy the standard conditions at infinity. Continuity of

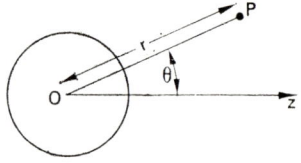

Fig. 5.3

V and $j_r (= \sigma E_r)$ at $r = a$ [see eqns. (5.13) and (5.15)] give

$$Aa\cos\theta = -E_0 a\cos\theta + \frac{B\cos\theta}{a^2},$$
$$-\sigma A\cos\theta = \sigma_0 \left(E_0 \cos\theta + \frac{2B\cos\theta}{a^3} \right).$$

Therefore $\quad a^3 A - B = -a^3 E_0, \quad \sigma a^3 A + 2\sigma_0 B = -\sigma_0 a^3 E_0.$
Therefore $\quad A = -3E_0\sigma_0/(2\sigma_0+\sigma), \quad B = a^3 E_0(\sigma-\sigma_0)/(2\sigma_0+\sigma).$

§ 5.7 THE STEADY FLOW OF ELECTRIC CURRENTS

Therefore, within the sphere
$$\mathbf{j} = \sigma \mathbf{E} = -\sigma\{0 \ 0 \ \partial V/\partial z\} = -\sigma\{0 \ 0 \ A\}$$
$$= \{0 \ 0 \ 3\sigma\sigma_0 E_0/(2\sigma_0+\sigma)\} = 3\sigma(2\sigma_0+\sigma)^{-1}\mathbf{j}_0$$

and, since this is constant, the flow through the sphere is uniform.

A special case. In the special case where $\sigma = 0$, e.g. when the material within the sphere $r = a$ has been removed, the only boundary condition at the surface of the sphere is $(j_r)_{r=a} = 0$ and we find, for $r \geqslant a$,

$$V = -E_0\left(r + \frac{a^3}{2r^2}\right)\cos\theta,$$

$$j_r = -\sigma_0\frac{\partial V}{\partial r} = \sigma_0 E_0\left(1 - \frac{a^3}{r^3}\right)\cos\theta, \quad j_\theta = -\sigma_0\frac{1}{r}\frac{\partial V}{\partial \theta} = \sigma_0 E_0\left(1 + \frac{a^3}{2r^3}\right)\sin\theta, \quad j_\psi = 0.$$

The differential equations for the lines of current flow (stream-lines), see C.M., vol. 4, § 4.2, are

$$\frac{dr}{j_r} = \frac{r\,d\theta}{j_\theta} = \frac{r\sin\theta\,d\psi}{j_\psi},$$

i.e. $\psi = $ constant,

and
$$\frac{dr}{\left(1 - \dfrac{a^3}{r^3}\right)\cos\theta} = \frac{r\,d\theta}{\left(1 + \dfrac{a^3}{2r^3}\right)\sin\theta}.$$

This last equation separates to give

$$\int \frac{(2r^3 + a^3)}{r(r^3 - a^3)}\,dr - \int 2\cot\theta\,d\theta = \text{constant}.$$

Integration, after using the substitution $r^3 = x$ in the first integral, gives the equations of the family of current flow lines as

$$(r^3 - a^3)\sin^2\theta = \lambda a^2 r, \quad \psi = \text{constant}$$

where λ is a dimensionless parameter. Since the system has axial symmetry we need only sketch the lines of flow in some plane $\psi = $ constant. They are shown in Fig. 5.4.

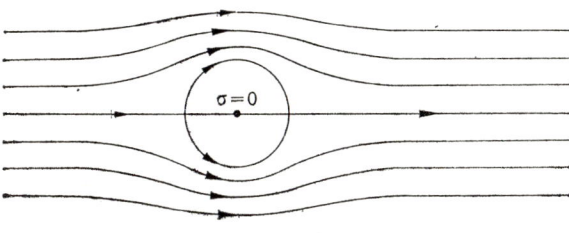

FIG. 5.4

Note that the semi-infinite lines $\theta = 0, r \geqslant a$, and the sphere $r = a$ form the stream surface $\lambda = 0$.

The results of this problem can also be interpreted in terms of the steady flow of incompressible uniform fluid past a rigid spherical barrier. [See the authors' *Elementary Classical Hydrodynamics*, p. 80, Pergamon Press.]

Example 3. Two small electrodes of radii δ_1, δ_2 are embedded in an infinite uniform medium of uniform conductivity σ with their centres at a distance b apart. Neglecting terms of order δ_1/b, δ_2/b, obtain an expression for the resistance between the electrodes.

First, we obtain an expression for the potential due to an isolated spherical electrode, of radius a, emitting total current I_0. Clearly from symmetry and the equation of continuity of charge,
$$4\pi r^2 j_r = I_0.$$
Therefore
$$j_r = \frac{I_0}{4\pi r^2} = -\sigma \frac{\partial V}{\partial r} \quad \text{so that} \quad V = \frac{I_0}{4\pi\sigma r},$$
the constant of integration being chosen so that V vanishes at infinity. Then the potential at the surface of the electrode is
$$V_0 = I_0/(4\pi\sigma a).$$

Suppose now that of the two given electrodes the one with centre A_1, Fig. 5.5, emits current I and the one with centre A_2 absorbs current I.

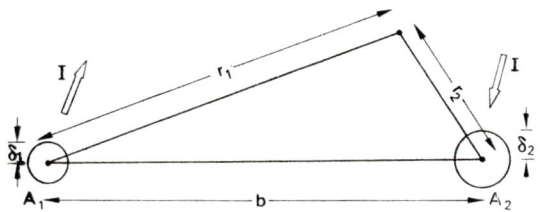

Fig. 5.5

We can find the potential as in the corresponding electrostatic problem of two charged spheres embedded in a uniform dielectric.

The potential near the surface of sphere A_2 due to the electrode A_1 must lie between
$$\frac{I}{4\pi\sigma(b+\delta_2)} \quad \text{and} \quad \frac{I}{4\pi\sigma(b-\delta_2)},$$
i.e.
$$\frac{I}{4\pi\sigma b}\left(1-\frac{\delta_2}{b}\right) \quad \text{and} \quad \frac{I}{4\pi\sigma b}\left(1+\frac{\delta_2}{b}\right).$$

But δ_2/b is negligible; hence the potential of A_2 due to A_1 is the constant $I/(4\pi\sigma b)$. Hence, to our degree of approximation the potential function
$$V = \frac{I}{4\pi\sigma}\left(\frac{1}{r_1}-\frac{1}{r_2}\right)$$
satisfies all the required conditions, in particular it corresponds to constant potentials on A_1 and A_2. These potentials are
$$V_1 = \frac{I}{4\pi\sigma}\left(\frac{1}{\delta_1}-\frac{1}{b}\right), \quad V_2 = \frac{I}{4\pi\sigma}\left(\frac{1}{b}-\frac{1}{\delta_2}\right).$$
Therefore
$$V_1 - V_2 = \frac{I}{4\pi\sigma}\left(\frac{1}{\delta_1}+\frac{1}{\delta_2}-\frac{2}{b}\right).$$
Therefore
$$R = \frac{V_1-V_2}{I} = \frac{1}{4\pi\sigma}\left(\frac{1}{\delta_1}+\frac{1}{\delta_2}-\frac{2}{b}\right).$$

§ 5.7 THE STEADY FLOW OF ELECTRIC CURRENTS

Example 4. The leaking cable.

A uniform cable of length l, whose resistance per unit length is ϱ_1, has an imperfectly insulating covering such that the resistance to earth is ϱ_2 per unit length. Prove that, when a steady current is flowing, the potential V satisfies the equation

$$\frac{d^2V}{dx^2} = n^2 V,$$

where x denotes the distance along the cable, and where $n^2 = \varrho_1/\varrho_2$.

Prove that, if one end of the cable is earthed, the ratio of the energies dissipated in the cable and in the insulation is

$$\frac{\sinh 2nl + 2nl}{\sinh 2nl - 2nl}.$$

Let the potential and current flowing along the cable at distances x, $x+\delta x$ from the end $x = 0$ be as shown in Fig. 5.6. Then, neglecting second and higher order terms, Ohm's law applied to the cable between normal cross-sections at A_1, A_2 gives

$$-\delta V = (\varrho_1 \, \delta x) I. \tag{1}$$

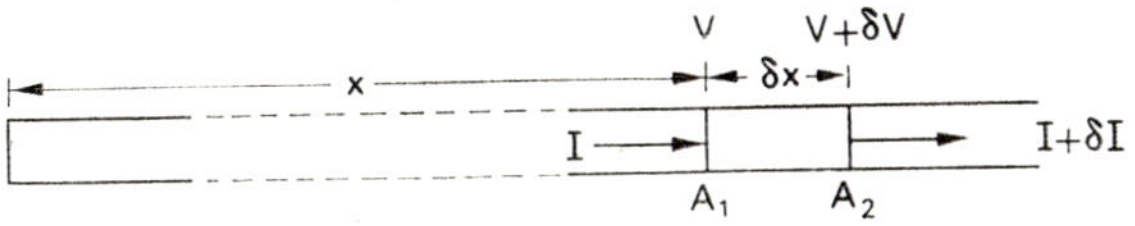

FIG. 5.6

We use the formula $\sigma = s/(\alpha R)$, see eqn. (5.6), to find the resistance between the segment $A_1 A_2$ and the earth. The length s is here the thickness of the insulation and the cross-section α is the product $p \, \delta x$, where p is the circumference of the cable and δx the length of the element. Therefore

$$R = s/(\alpha p \, \delta x) = k/\delta x,$$

i.e. the resistance between the element and the earth is *inversely proportional* to the length of the element. Hence for the element $A_1 A_2$ the current $-\delta I$ which leaks to earth flows through a resistance $\varrho_2/\delta x$ under a potential difference V. Therefore

$$V = -\frac{\varrho_2 \, \delta I}{\delta x}. \tag{2}$$

(Alternatively, we see that the elements are connected to earth in *parallel* and so we use the reciprocal formula of p. 175 for the equivalent resistance of resistors in parallel.)

Letting $\delta x \to 0$, eqns. (1) and (2) give

$$\frac{dV}{dx} = -\varrho_1 I, \quad V = -\varrho_2 \frac{dI}{dx}. \tag{3}$$

Assuming ϱ_1, ϱ_2 are constant, elimination of I gives

$$\frac{d^2V}{dx^2} = n^2 V, \quad n^2 = \frac{\varrho_1}{\varrho_2}. \tag{4}$$

Let V_0 be the potential at the end $x = 0$. Then the solution of eqn. (4) which satisfies the boundary conditions $V = V_0$ at $x = 0$, $V = 0$ at $x = l$ is

$$V = \frac{V_0 \sinh\{n(l-x)\}}{\sinh nl}. \tag{5}$$

The rate of dissipation of energy in the element A_1A_2 is $\varrho_1 I^2 \delta x$. Therefore the total rate of dissipation of energy in the cable is

$$D_1 = \int_0^l \varrho_1 I^2 \, dx = \int_0^l \frac{1}{\varrho_1}\left(\frac{dV}{dx}\right)^2 dx = \int_0^l \frac{V_0^2 n^2 \cosh^2\{n(l-x)\} \, dx}{\varrho_1 \sinh^2 nl}$$

$$= \frac{V_0^2 n^2}{2\varrho_1 \sinh^2 nl} \int_0^l [1+\cosh\{2n(l-x)\}] \, dx,$$

i.e.
$$D_1 = V_0^2 n(2nl + \sinh 2nl)/\{4\varrho_1 \sinh^2 nl\}. \tag{6}$$

From eqn. (2) the rate of dissipation of energy in the insulator surrounding the segment A_1A_2 is $V^2 \delta x/\varrho_2$ since the resistance of this insulation is $\varrho_2/\delta x$. Therefore

$$D_2 = \int_0^l \frac{V^2 \, dx}{\varrho_2} = \frac{V_0^2}{\varrho_2 \sinh^2 nl} \int_0^l \sinh^2\{n(l-x)\} \, dx = \frac{V_0^2 (\sinh 2nl - 2nl)}{4\varrho_2 n \sinh^2 nl}.$$

Therefore
$$\frac{D_1}{D_2} = \frac{n^2 \varrho_1}{\varrho_2} \cdot \frac{\sinh 2nl + 2nl}{\sinh 2nl - 2nl} = \frac{\sinh 2nl + 2nl}{\sinh 2nl - 2nl}$$

as required.

According to these results total dissipation of heat takes place at a rate

$$D_1 + D_2 = \frac{V_0^2}{4\sinh^2 nl} \left\{\frac{n}{\varrho_1}(\sinh 2nl + 2nl) + \frac{1}{n\varrho_2}(\sinh 2nl - 2nl)\right\}$$

$$= \frac{V_0^2 n}{4\varrho_1 \sinh^2 nl} \cdot 2\sinh 2nl = \frac{V_0^2 n}{\varrho_1} \coth nl.$$

From the first of eqns. (3) the current is

$$I = -\frac{1}{\varrho_1}\frac{dV}{dx} = \frac{V_0 n}{\varrho_1} \cdot \frac{\cosh\{n(l-x)\}}{\sinh nl}.$$

Hence the current I_0 entering the cable at $x = 0$ is

$$I_0 = \frac{V_0 n}{\varrho_1} \coth nl,$$

and, since the potential there is V_0, the rate of supply of energy to the system is

$$I_0 V_0 = (V_0^2 n/\varrho_1)\coth nl = D_1 + D_2$$

illustrating the general principle of conservation of energy.

Exercises 5.7

1. A large slab of material with parallel plane faces and thickness a has specific conductivity σ which varies uniformly from a value σ_1 at one face to σ_2 at the other. If the faces are maintained at potentials V_1 and V_2 respectively, prove that the steady current j per unit area which flows between them is

$$j = \frac{V_1 - V_2}{a} \cdot \frac{\sigma_1 - \sigma_2}{\ln(\sigma_1/\sigma_2)}.$$

If also the dielectric constant K varies uniformly from K_1 at the first face to K_2 at the second, prove that at any point of the material there is a space-charge of density

$$\varepsilon_0 j(K_2 \sigma_1 - K_1 \sigma_2)/(a\sigma^2).$$

§ 5.7 THE STEADY FLOW OF ELECTRIC CURRENTS

2. A block of conducting material is in the form of a cube of side a, and the conductivity at any point in the block is $\lambda(a+x)$, where x is the distance of the point from a face S of the cube and λ is a constant. Show that the resistance of the block between S and the opposite face is $\dfrac{1}{\lambda a^2} \ln 2$, and find the resistance between a pair of opposite faces perpendicular to S. (It may be assumed that the currents are unidirectional in each case.)

3. A and B are opposite ends of a diameter of a thin spherical shell of centre O, radius a and thickness t. Current enters and leaves by two small circular electrodes of radius $c\,(\gg t)$ whose centres are at A and B. Show that the potential V at a point of the shell satisfies the equation
$$\frac{\partial}{\partial \theta}\left(\sin\theta\,\frac{\partial V}{\partial \theta}\right) = 0.$$

If i is the total current and P is a point on the shell such that the angle $POA = \theta$, show that the magnitude of the current vector at P is $i/(2\pi at\sin\theta)$. Deduce that the resistance of the conductor is approximately $\dfrac{1}{\pi\sigma t}\ln\dfrac{2a}{c}$, where σ is the conductivity.

4. A steady current flows through a homogeneous substance bounded by two infinite cylindrical electrodes whose cross-sections are concentric circles of radii a and $b\,(b>a)$. The conductivity σ at a distance r from the axis of the cylinders is given by $\sigma_0 a/r$, where σ_0 is constant. Find the potential V at any point of the substance, and show that the resistance R per unit length is given by
$$2\pi a\sigma_0 R = b-a.$$

5. A steady current of constant density C flows parallel to a fixed direction in a uniform infinite plane conducting sheet of constant conductivity σ. A circular portion of radius a is replaced by one of constant conducitivty σ'. Show that the current flowing through this portion is
$$4a\sigma'C/(\sigma+\sigma').$$

Sketch the current lines in the cases where (i) σ' is finite and greater than σ, (ii) σ' is infinite.

6. A plane sheet of metal of uniform thickness t and conductivity σ has a boundary $ABCDEFA$, composed of two concentric semicircular arcs ABC, DEF and two straight edges CD, FA in the same straight line, the semicircles being on the same side of the line $FACD$. When the edges CD, FA are maintained at a constant potential difference show that the resistance of the plate is
$$\frac{\pi}{\sigma t\ln(b/a)},$$
where a and b are respectively the radii of the inner and outer semicircles.
Also find the resistance of the plate when the semicircles are maintained at a constant potential difference.

7. An underground cable consists of a conducting core surrounded by imperfect insulating material. The resistance of a unit length of the core is R, and the resistance of the insulating material in a unit length of the cable to a current flowing from the core to earth is r. Obtain the equations
$$\frac{dV}{dx} = -Ri, \qquad \frac{di}{dx} = -\frac{V}{r},$$
where V is the potential of the core and i is the current flowing along the core at a distance x from one end.
Find, in terms of R, r and the length l of the cable the resistance to a current entering at one end (i) when the other end is earthed, (ii) when the other end is insulated.

5.8 Integral theorems

Because of the analogy with electrostatics there exist theorems applicable to current flow similar to those for electrostatics; we give below the appropriate forms of these theorems.

1. The Uniqueness Theorem

This theorem states that there is only one possible distribution of current in a medium of prescribed conductivity σ and which corresponds to given electrodes and electromotive forces; the current flow analogue of the electrostatic case is that *either* the potential *or* the current strength of each electrode must be prescribed to determine the current flow uniquely; both need not (in fact cannot) be specified. The proof of the theorem uses Kelvin's generalization of Green's theorem, following closely the lines of the proof given in § 4.3.

We suppose that there are two possible functions V, V' corresponding to current distributions j, j' which satisfy all the conditions:

(i) $j = \sigma(E+\mathcal{E})$, div $j = 0$; $j' = \sigma(E'+\mathcal{E})$, div $j' = 0$.

Since the electromotive field is specified, $\mathcal{E}$ is the same in both distributions.

(ii) $E = -\operatorname{grad} V$; $E' = -\operatorname{grad} V'$.

(iii) At all discontinuities

$$V_+ - V_- = \psi; \quad V'_+ - V'_- = \psi.$$

(iv) On each electrode V and V' are constant, but on some S_i, $V = V_i = V'$ when the potential is prescribed, and on the remainder of the electrodes S_j,

$$\iint_{S_j} j \cdot dS = \iint_{S_j} \sigma \frac{\partial V}{\partial \nu} dS = I_j = \iint_{S_j} \sigma \frac{\partial V'}{\partial \nu} dS = \iint_{S_j} j' \cdot dS.$$

(v) On the free boundary, i.e. that part of the boundary which is not an electrode,

$$j' \cdot \hat{\nu} = j \cdot \hat{\nu}.$$

If we write $U = V' - V$ we see that, at all points in the conductor div ($\sigma \operatorname{grad} U$) = 0, and at all surfaces of discontinuity U is continuous, and on each electrode

$$\text{either} \quad U = 0, \quad \text{or} \quad \iint \sigma \frac{\partial U}{\partial \nu} dS = 0.$$

Also at a free boundary $\partial U/\partial n = 0$.

We apply the theorem

$$\iiint \{U \operatorname{div} (\sigma \operatorname{\mathbf{grad}} U) + \sigma (\operatorname{\mathbf{grad}} U)^2\} \, d\tau = \iint \sigma U \frac{\partial U}{\partial n} \, dS$$

to the region bounded by the electrodes and the free boundary (and a large surface if the conducting region extends to infinity, in which case we assume that both V, V' and therefore U, satisfy the standard boundary conditions). The conditions which apply at the electrodes (and at infinity) imply that

$$\iint \sigma U \frac{\partial U}{\partial n} \, dS = 0,$$

where $\partial/\partial n$ denotes differentiation along the outward normal of the region of integration. Hence

$$\iiint \sigma (\operatorname{\mathbf{grad}} U)^2 \, d\tau = 0.$$

Therefore $U = $ constant, the constant being zero in practically all circumstances, and the two distributions $\mathbf{j}$, $\mathbf{j}'$ are identical.

2. The Reciprocal Theorem

A given geometrical arrangement of electrodes in a conductor can produce a number of different current distributions according to the strengths and potentials of the different electrodes. Suppose that a current distribution $\mathbf{j}$ corresponds to electrodes S_r at potentials V_r and strengths I_r, and that a second distribution $\mathbf{j}'$ corresponds to V'_r, I'_r. From Kelvin's generalization of another form of Green's theorem we have

$$\iiint \{\operatorname{div}(\sigma V \operatorname{\mathbf{grad}} V') - \operatorname{div}(\sigma V' \operatorname{\mathbf{grad}} V)\} \, d\tau = \iint \sigma \left(V \frac{\partial V'}{\partial n} - V' \frac{\partial V}{\partial n} \right) dS$$

where the right-hand side is taken over the boundary of the volume and $\partial/\partial n$ denotes differentiation along the outward normal. When there is no source of e.m.f. within the conducting volume the left-hand side of this equation vanishes. The boundary of the conducting region consists either of insulating areas, where $\partial V/\partial n$, $\partial V'/\partial n$ vanish, or electrodes on which V, V' take constant values. Therefore

$$\sum_r \iint_{S_r} \sigma V \frac{\partial V'}{\partial n} \, dS = \sum_r \iint_{S_r} \sigma V' \frac{\partial V}{\partial n} \, dS.$$

But on $\quad S_r, V = V_r, V' = V'_r, I_r = -\iint_{S_r} \sigma \dfrac{\partial V}{\partial n}\,dS, I'_r = -\iint_{S_r} \sigma \dfrac{\partial V'}{\partial n}\,dS.$

Therefore
$$\sum_r V_r I'_r = \sum_r V'_r I_r. \qquad (5.20)$$

3. The heating effect

Energy is supplied to a system at the rate $V_r I_r$ when a current of strength I_r enters by an electrode at potential V_r; if there is an e.m.f. ψ maintained across a surface K, it supplies energy at a rate $\iint_K \psi \mathbf{j}\cdot d\mathbf{S}$, and an electromotive field $\mathcal{E}$ supplies energy at the rate $\iiint \mathbf{j}\cdot\mathcal{E}\,d\tau$. Since the state of the system is steady this energy appears as heat produced at the rate

$$H = \iiint \mathbf{j}\cdot\mathcal{E}\,d\tau + \iint_K \psi \mathbf{j}\cdot d\mathbf{S} + \sum_r I_r V_r.$$

The vector $\mathbf{j}$ satisfies the conditions

$$\operatorname{div} \mathbf{j} = 0, \quad \mathbf{j} = \sigma(\mathbf{E}+\mathcal{E}), \quad \mathbf{E} = -\operatorname{\mathbf{grad}} V.$$

Therefore

$$\iiint \mathbf{j}\cdot\mathcal{E}\,d\tau = \iiint \mathbf{j}\cdot(\mathbf{j}/\sigma + \operatorname{\mathbf{grad}} V)\,d\tau$$
$$= \iiint (j^2/\sigma)\,d\tau + \iiint \mathbf{j}\cdot\operatorname{\mathbf{grad}} V\,d\tau$$
$$= \iiint (j^2/\sigma)\,d\tau + \iiint \{\operatorname{div}(V\mathbf{j}) - V\operatorname{div} \mathbf{j}\}\,d\tau.$$

We transform the integral $\iiint \operatorname{div}(V\mathbf{j})\,d\tau$ to give surface integrals over the bounding electrodes S_r or over an envelope S_K enclosing the surface K carrying the e.m.f. ψ, thus

$$\iiint \operatorname{div}(V\mathbf{j})\,d\tau = \sum_r \iint_{S_r} V(\mathbf{j}\cdot\hat{\mathbf{n}})\,dS + \iint_{S_k} V(\mathbf{j}\cdot\hat{\mathbf{n}})\,dS.$$

On each electrode $V = V_r$, $\iint_{S_r}(\mathbf{j}\cdot\hat{\mathbf{n}})\,dS = -I_r$. The relation between the

§ 5.8   THE STEADY FLOW OF ELECTRIC CURRENTS   169

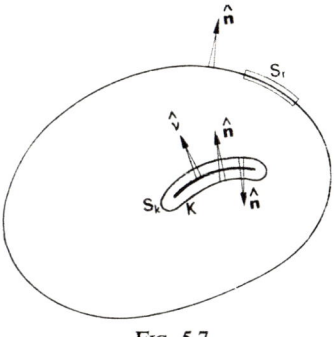

Fig. 5.7

unit normal $\hat{\boldsymbol{\nu}}$ of the surface K, and the outward normals $\hat{\boldsymbol{n}}$ of the volume of integration are shown in Fig. 5.7. Hence, in the limit as S_k tends to coincide with K,

$$\iint_{S_k} V(\boldsymbol{j}\cdot\hat{\boldsymbol{n}})\,\mathrm{d}S \to \iint_K (V_- - V_+)(\boldsymbol{j}\cdot\hat{\boldsymbol{n}})\,\mathrm{d}S = -\iint_K \psi\boldsymbol{j}\cdot\mathrm{d}\boldsymbol{S}.$$

Therefore $$\iiint (\boldsymbol{j}\cdot\boldsymbol{\mathscr{E}})\,\mathrm{d}\tau = \iiint (j^2/\sigma)\,\mathrm{d}\tau - \sum_r V_r I_r - \iint_K \psi\boldsymbol{j}\cdot\mathrm{d}\boldsymbol{S}.$$

Therefore $$H = \iiint (j^2/\sigma)\,\mathrm{d}\tau. \tag{5.21}$$

4. Minimum theorems

The theorem we prove now is analogous to Kelvin's minimum energy theorem in electrostatics. We consider a steady current flow related to a given set of electrodes and electromotive forces; the actual flow is $\boldsymbol{j}$ and an arbitrary flow related to the same sources is $\boldsymbol{j}+\boldsymbol{j}'$, where

$$\mathrm{div}\,\boldsymbol{j} = 0 = \mathrm{div}\,(\boldsymbol{j}+\boldsymbol{j}'), \quad \text{i.e.} \quad \mathrm{div}\,\boldsymbol{j}' = 0. \tag{5.22}$$

On electrodes $$\iint_{S_r} \boldsymbol{j}\cdot\mathrm{d}\boldsymbol{S} = I_r = \iint_{S_r} (\boldsymbol{j}+\boldsymbol{j}')\cdot\mathrm{d}\boldsymbol{S}, \quad \text{i.e.} \quad \iint_{S_r} \boldsymbol{j}'\cdot\mathrm{d}\boldsymbol{S} = 0, \tag{5.23}$$

in the field $\quad \boldsymbol{j} = \sigma(\boldsymbol{E}+\boldsymbol{\mathscr{E}}), \quad \boldsymbol{j}+\boldsymbol{j}' = \sigma(\boldsymbol{E}+\boldsymbol{E}'+\boldsymbol{\mathscr{E}}), \quad \text{i.e.} \quad \boldsymbol{j}' = \sigma\boldsymbol{E}'. \tag{5.24}$

[The relation (5.24) is slightly more general than that assumed for the corresponding electrostatic condition—it would correspond to a prescribed distribution of "permanent" polarization.]

From eqns. (5.22) and (5.24) we obtain div $(\sigma E') = 0$, and the method, identical with that in § 4.5, shows that the integral

$$I = \iiint \sigma E^2 \, d\tau$$

has a minimum value when the following conditions are satisfied

curl $E = 0$ in the field,
$\hat{\nu} \times E = 0$ on all electrodes,
$\hat{\nu} \times (E_+ - E_-) = 0$ on all surfaces of discontinuity.

These are the conditions satisfied by E in the actual distribution j and correspond to the existence of a potential function V which may have some discontinuities, $V_+ - V_- = \psi$ on K, and is constant on all electrodes. Since $E = (j/\sigma - \mathcal{E})$,

$$I = \iiint \{(j^2/\sigma) - 2j \cdot \mathcal{E} + \sigma \mathcal{E}^2\} \, d\tau$$

and because $\mathcal{E}$ is the given distribution of the electromotive field we deduce that the actual current distribution is that which makes the integral

$$H = \iiint \{(j^2/\sigma) - 2j \cdot \mathcal{E}\} \, d\tau \tag{5.25}$$

a minimum. For the special case in which there is no electromotive field, and the distribution arises solely from electrodes, $\mathcal{E} = 0$, and eqn. (5.25) implies that the actual distribution of current is that which makes

$$H = \iiint (j^2/\sigma) \, d\tau \tag{5.26}$$

a minimum, i.e. the rate of heat production is a minimum.

When there are surfaces across which an e.m.f. acts, the condition $\hat{\nu} \times (E_+ - E_-) = 0$ implies that the potential has a fixed discontinuity $\psi = V_+ - V_-$ across each such surface and the second term in (5.25) is replaced by the limiting case when $\mathcal{E}$ becomes a discontinuity so that

$$\iiint (j \cdot \mathcal{E}) \, d\tau = \psi \iint (j \cdot \hat{\nu}) \, dS = \psi i.$$

Hence the distribution of current is such as to make

$$\iiint (j^2/\sigma) \, d\tau - 2 \sum \psi i$$

a minimum, where the sum is extended over all surfaces carrying e.m.f. through which the current i passes.

§ 5.8 THE STEADY FLOW OF ELECTRIC CURRENTS

Example. Two electrodes A and B are embedded in a conductor which is free from all other sources of current or e.m.f. If the conductivity is increased, the total resistance between A and B decreases.

This problem is the current analogue of example 2, p. 134 of Chapter 4 in which the dielectric constant is increased. The conditions there are more elaborate than those considered here.

We suppose that the current I enters at A and leaves at B, then the resistance is given by $IR = V_A - V_B$ and

$$I = \iint_A \sigma \frac{\partial V}{\partial n} \, dS = -\iint_B \sigma \frac{\partial V}{\partial n} \, dS,$$

where $\partial/\partial n$ is directed out of the conducting medium. We give two cases.

(a) The conductivity is increased whereas I is kept constant.

Since
$$H = \iiint \sigma \, (\text{grad } V)^2 \, d\tau = I^2 R = I(V_A - V_B),$$

$$\delta H = I^2 \delta R = \iiint \delta\sigma \, (\text{grad } V)^2 \, d\tau + 2 \iiint \sigma \, \text{grad } V \cdot \text{grad } (\delta V) \, d\tau.$$

But $\iiint \sigma \, \text{grad } V \cdot \text{grad } \delta V \, d\tau = -\iint \delta V \, \text{div} \, (\sigma \, \text{grad } V) \, d\tau + \iint_{A+B} \delta V \sigma \frac{\partial V}{\partial n} \, dS.$

On the electrodes the potential is constant, and in the conductor $\text{div}\,(\sigma \, \text{grad } V) = 0$. Therefore

$$\iiint \sigma \, \text{grad } V \cdot \text{grad } \delta V \, d\tau = (\delta V_A - \delta V_B)I = \delta H.$$

Therefore
$$\delta H = \iiint \delta\sigma \, (\text{grad } V)^2 \, d\tau + 2\delta H.$$

Therefore
$$\delta H = I^2 \, \delta R = -\iiint \delta\sigma \, (\text{grad } V)^2 \, d\tau.$$

Since we assumed that $\delta\sigma \geqslant 0$ we conclude that $\delta R \leqslant 0$.

(b) The conductivity is increased whereas the potentials V_A, V_B of the electrodes are kept constant. In this case

$$H = \iiint \sigma \, (\text{grad } V)^2 \, d\tau = (V_A - V_B)^2 / R.$$

Therefore
$$\delta H = -\{(V_A - V_B)^2 / R^2\} \, \delta R = \iiint \delta\sigma \, (\text{grad } V)^2 \, d\tau + \iint_{A+B} \delta V \sigma \frac{\partial V}{\partial n} \, dS.$$

The surface integrals vanish since we postulated that V_A and V_B are fixed. Therefore

$$\delta H = -\{(V_A - V_B)^2 / R^2\} \, \delta R = \iiint \delta\sigma \, (\text{grad } V)^2 \, d\tau.$$

Again, if $\delta\sigma \geqslant 0$ we conclude that $\delta R \leqslant 0$.

(The reader should compare the circumstances of this problem with the similar arrangements in § 3.4, p. 100, where displacements at constant potential were considered.)

5.9 Networks of linear conductors

A network of conducting wires through which flows electric current is a common example of a current distribution and forms an important special case of the general distribution. The conducting wires form a (multiply connected) region to which we can apply the theorems already established.

We consider a network of wires joining points (or nodes) $A, B, C, \ldots$, in which the potentials of the nodes are $V_A, V_B, V_C, \ldots$, the resistances of the wires joining $AB, AC, BC, \ldots, R_{AB}, R_{AC}, R_{BC}, \ldots$, and the corresponding currents are $i_{AB} = -i_{BA}, i_{AC} = -i_{CA}, i_{BC} = -i_{CB}$, etc. If there is a source of e.m.f. in the wire joining two points, we denote it by $\psi_{AB} = -\psi_{BA}, \psi_{AC} = -\psi_{CA}$,

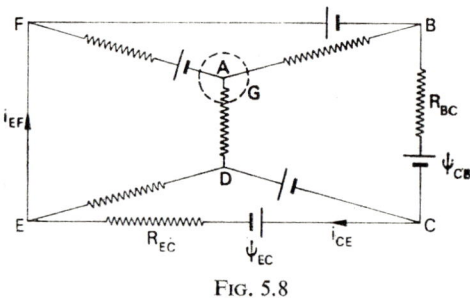

Fig. 5.8

$\psi_{BC} = -\psi_{CB}$, etc. Diagrammatically such a network is represented in Fig. 5.8. The longer of the two lines representing an e.m.f. gives the high potential side of the discontinuity and the order of the letters, e.g. ψ_{CB}, indicates that, when ψ_{CB} is positive, the e.m.f. tends to send a current from C to B across the discontinuity; the longer line is the "positive" terminal of the battery. The other symbols are self-explanatory.

Kirchhoff's laws

The two laws associated with the name of Kirchhoff are the special cases of eqns. (5.4) and (5.6), or the more general eqn. (5.19), applied to a network.

The first law is obtained from eqn. (5.4), div $\boldsymbol{j} = 0$, by drawing an arbitrary closed surface G enclosing a node (see Fig. 5.8). Then

$$\iiint_G \operatorname{div} \boldsymbol{j} \, d\tau = \oiint_G \boldsymbol{j} \cdot d\boldsymbol{S} = 0.$$

But, on $G, \boldsymbol{j}$ is different from zero only where the wires connected to the node

§ 5.9 THE STEADY FLOW OF ELECTRIC CURRENTS

$$\oint_G \mathbf{j} \cdot \mathbf{dS} = \sum_K i_{AK} = 0. \tag{5.27}$$

A cut it. Therefore

In words, the sum of the currents leaving A (or any node) is zero.

The second of Kirchhoff's laws is obtained by integrating eqn. (5.19) around a closed contour.

$$\oint (\mathbf{j}/\sigma - \mathcal{E}) \cdot \mathbf{ds} = \oint \mathbf{E} \cdot \mathbf{ds} = 0,$$

since **curl $E = 0$**. We choose as the path of integration any closed loop in the network (e.g. $ABCDA$); then

$$\oint_{ABCDA} = \int_A^B + \int_B^C + \int_C^D + \int_D^A.$$

In a segment such as BC,

$$\mathbf{j} = \frac{i_{BC}}{\alpha} \hat{\mathbf{e}}, \quad \int_B^C \mathbf{j} \cdot \mathbf{ds} = i_{BC} \int_B^C \frac{\hat{\mathbf{e}} \cdot \mathbf{ds}}{\sigma \alpha} = i_{BC} R_{BC},$$

where $\hat{\mathbf{e}}$ is a unit vector along the wire. Also

$$\int_B^C \mathcal{E} \cdot \mathbf{ds} = \psi_{BC} = -\psi_{CB}.$$

Hence, by adding the integrals for each segment of the loop we obtain

$$\sum_{\text{loop}} (i_{AB} R_{AB} - \psi_{AB}) = 0. \tag{5.28}$$

This is Kirchhoff's second law.

The use of Kirchhoff's first two laws to determine the currents flowing in a network can be facilitated by the use of circulatory currents. Since div $\mathbf{j} = 0$ it follows that the tubes of flow are either closed loops or begin and end on electrodes. When applied to a network this means that Kirchhoff's first law is satisfied identically by assuming arbitrary currents circulating in closed loops of the network and other currents (if required) which enter at one point and leave at another, i.e. at electrodes, and follow an arbitrary path through the network. Application of Kirchhoff's second law to all the independent closed loops of the circuit then determines the magnitudes of these arbitrary currents uniquely. Success in easily solving such problems depends upon the right choice of loops to be used; the reader will find these points illustrated in the examples which follow.

Example 1. In the closed network of Fig. 5.9 (i) there are two independent loops and we choose circulating currents i_1 and i_2 in these loops *ABD, BCD* as shown in Fig. 5.9 (i). The currents in the various branches of the network implied by these assumptions are indicated in Fig. 5.9 (ii). Because no current enters or leaves the network from outside we need assume no other currents to be present, and obviously Kirchhoff's first law is satisfied for every node. We now apply the second law to any convenient pair of (independent) loops.

Loop *ABDA*: $\quad\quad i_1 R_{AB} + (i_1 - i_2) R_{BD} + i_1 R_{DA} = E,$

loop *DBCD*: $\quad\quad i_2 R_{DC} + i_2 R_{BC} - (i_1 - i_2) R_{BD} = -E.$

We can determine both i_1 and i_2 from these equations.

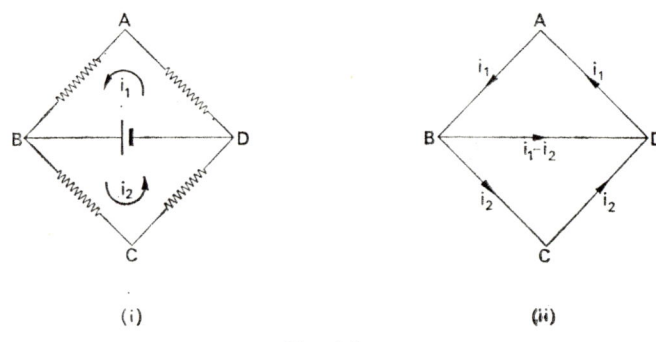

Fig. 5.9

Example 2. The resistance of n (resistance) wires in series.

Figure 5.10 shows n wires $A_1 B_1, A_2 B_2, \ldots, A_n B_n$ of resistances $R_1, R_2, \ldots, R_n$ respectively connected in series, i.e. in a line so that B_1 is connected to A_2, B_2 to A_3, etc. The standard symbol for representing a resistance is ⟿. Suppose that the potentials at $A_1, A_2, \ldots, A_n, B_n$ are $V_1, V_2, \ldots, V_n, V_{n+1}$ respectively and that a current i flows along the wire from A_1 to B_n. Then Ohm's law applied to each of the wires gives

$$V_1 - V_2 = iR_1, \quad V_2 - V_3 = iR_2, \ldots, V_n - V_{n+1} = iR_n.$$

By addition

$$V_1 - V_{n+1} = i(R_1 + R_2 + \ldots + R_n). \tag{1}$$

But by definition the equivalent resistance between A_1 and B_n of the n wires is R where

$$V_1 - V_{n+1} = iR. \tag{2}$$

It follows from eqns. (1) and (2) that

$$R = R_1 + R_2 + \ldots + R_n.$$

Fig. 5.10

Example 3. The resistance of n wires in parallel.

Suppose now that the n (resistance) wires of example 2 are connected in parallel, i.e. the ends $A_1, A_2, \ldots, A_n$ are connected by wires of negligible resistance to the point A and the ends $B_1, B_2, \ldots, B_n$ are similarly connected to the point B, Fig. 5.11, the potentials of A, B being V_A, V_B respectively. Then if $i_1, i_2, \ldots, i_n$ are the currents flowing in A_1B_1, $A_2B_2, \ldots, A_nB_n$ respectively, Kirchhoff's first law gives the total current flowing as

$$i' = i_1 + i_2 + \ldots + i_n. \tag{1}$$

Ohm's law applied to each of the resistances $A_1B_1, A_2B_2, \ldots, A_nB_n$ gives

$$i_1 R_1 = i_2 R_2 = \ldots = i_n R_n = V_A - V_B. \tag{2}$$

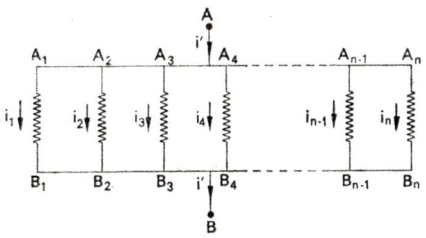

Fig. 5.11

From eqns. (1) and (2)

$$i' = i_1 + i_2 + \ldots + i_n = (V_A - V_B)\left(\frac{1}{R_1} + \frac{1}{R_2} + \ldots + \frac{1}{R_n}\right).$$

But the equivalent resistance R' is defined by $i' = (V_A - V_B)/R'$. Therefore

$$\frac{1}{R'} = \frac{1}{R_1} + \frac{1}{R_2} + \ldots + \frac{1}{R_n}.$$

Note. Instead of applying Ohm's law directly as above, we can frame the solution in terms of Kirchhoff's second law as follows:

loop $\quad AA_1B_1BB_2A_2A: \quad i_1R_1 - i_2R_2 = 0,$

loop $\quad AA_2B_2BB_3A_3A: \quad i_2R_2 - i_3R_3 = 0,$

$\quad\quad\quad\quad \ldots \quad\quad\quad\quad\quad\quad\quad \ldots$

loop $\quad AA_{n-1}B_{n-1}BB_nA_nA: \quad i_{n-1}R_{n-1} - i_nR_n = 0.$

These are equivalent to eqns. (2).

The results of examples 2 and 3 should be compared with those of example 6 on p. 89, for condensers. It should be noted that in illustration of the general result of p. 159, these results are the same when the capacity C of a condenser is replaced by the reciprocal $1/R$ of the resistance of a wire.

Example 4. A cube $ABCDEFGH$ is made of twelve equal wires each of resistance r. A current enters at one corner A and leaves by the opposite corner G. Find the equivalent resistance of the cube between the corners.

Using Kirchhoff's first law and the symmetry of the figure the currents in the wires are as shown in Fig. 5.12, the total current flowing being $3i$. Ohm's law applied by any path join-

176 ELEMENTARY ELECTROMAGNETIC THEORY

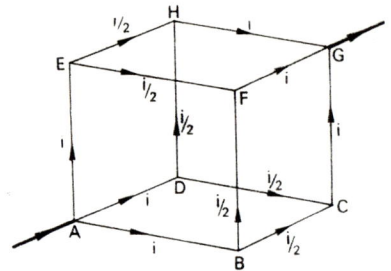

Fig. 5.12

ing A and G gives
$$V_A - V_G = 5ir/2.$$

The equivalent resistance R of the cube between A and G is defined by $V_A - V_G = 3iR$, and so
$$R = 5r/6.$$

Example 5. A tetrahedron $ABCD$ consists of conducting wires such that the resistances of pairs of opposite edges are equal. The resistances of the edges AD, BD and CD are R/λ R/μ and R, respectively. Determine the effective resistance of the terahedron when current enters at C and leaves at D.

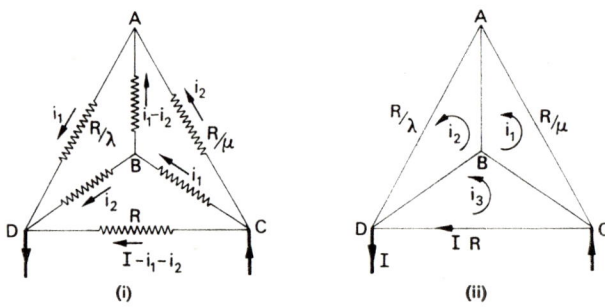

Fig. 5.13

Suppose that a current I enters at C and leaves at D [Fig. 5.13 (i)] and let the currents in CB, CA be i_1, i_2 respectively as shown. Then the symmetry of the figure implies that currents i_1, i_2 flow in AD, BD respectively. Kirchhoff's first law then implies that currents $I-i_1-i_2$, i_1-i_2 flow in CD, BA respectively.

Let V be the potential difference between C and D. Then applying the second law we obtain the following results.

$$CD: \quad R(I-i_1-i_2) = V; \tag{1}$$

$$CBD: \quad \frac{R}{\lambda} i_1 + \frac{R}{\mu} i_2 = V; \tag{2}$$

$$CBAC: \quad \frac{R}{\lambda} i_1 + R(i_1-i_2) - \frac{R}{\mu} i_2 = 0. \tag{3}$$

§ 5.9   THE STEADY FLOW OF ELECTRIC CURRENTS   177

Equation (3) gives $i_1\left(\dfrac{1}{\lambda}+1\right) = i_2\left(\dfrac{1}{\mu}+1\right) = \alpha$ (say).

Then eqn. (2) gives $\alpha = V\Big/\left\{R\left(\dfrac{1}{\lambda+1}+\dfrac{1}{\mu+1}\right)\right\}$

and eqn. (1) gives $I = \dfrac{V}{R} + \alpha\left(\dfrac{\lambda}{\lambda+1}+\dfrac{\mu}{\mu+1}\right) = \dfrac{V}{R}\left(1+\dfrac{\lambda/(\lambda+1)+\mu/(\mu+1)}{1/(\lambda+1)+1/(\mu+1)}\right)$

$= \dfrac{2V}{R}\left(\dfrac{1}{\lambda+1}+\dfrac{1}{\mu+1}\right)^{-1}.$

But, if r is the effective resistance of the network, by definition $I = V/r$. Therefore

$$r = \dfrac{R}{2}\left(\dfrac{1}{\lambda+1}+\dfrac{1}{\mu+1}\right).$$

Alternatively. Take i_1, i_2, i_3 to be the circulations as shown in Fig. 5.13 (ii) and suppose that an additional current I flows along CD.

Kirchhoff's second law applied to the specified loops gives:

$ABC:\quad i_1\left(1+\dfrac{1}{\lambda}+\dfrac{1}{\mu}\right)-i_2-\dfrac{i_3}{\lambda}=0;$ \hfill (4)

$ADB:\quad -i_1+i_2\left(1+\dfrac{1}{\lambda}+\dfrac{1}{\mu}\right)-\dfrac{i_3}{\mu}=0;$ \hfill (5)

$BDC:\quad -\dfrac{i_1}{\lambda}-\dfrac{i_2}{\mu}+i_3\left(1+\dfrac{1}{\lambda}+\dfrac{1}{\mu}\right)=I.$ \hfill (6)

From eqns. (4) and (5) we find

$$\dfrac{i_1}{\left(\dfrac{1}{\lambda}+\dfrac{1}{\mu}\right)\left(1+\dfrac{1}{\lambda}\right)} = \dfrac{i_2}{\left(\dfrac{1}{\lambda}+\dfrac{1}{\mu}\right)\left(1+\dfrac{1}{\mu}\right)} = \dfrac{i_3}{\left(2+\dfrac{1}{\lambda}+\dfrac{1}{\mu}\right)\left(\dfrac{1}{\lambda}+\dfrac{1}{\mu}\right)},$$

i.e. $\dfrac{i_1}{1+\dfrac{1}{\lambda}} = \dfrac{i_2}{1+\dfrac{1}{\mu}} = \dfrac{i_3}{2+\dfrac{1}{\lambda}+\dfrac{1}{\mu}} = k$ (say).

Substituting into eqn. (6),

$$k\left\{-\dfrac{1}{\lambda}-\dfrac{1}{\lambda^2}-\dfrac{1}{\mu}-\dfrac{1}{\mu^2}+2+3\left(\dfrac{1}{\lambda}+\dfrac{1}{\mu}\right)+\left(\dfrac{1}{\lambda}+\dfrac{1}{\mu}\right)^2\right\} = I,$$

i.e. $k = \dfrac{I}{2\left\{1+\dfrac{1}{\lambda}+\dfrac{1}{\mu}+\dfrac{1}{\lambda\mu}\right\}} = \dfrac{I\lambda\mu}{2(1+\lambda)(1+\mu)}.$

Therefore $i_3 = \dfrac{(\lambda+\mu+2\lambda\mu)}{2(1+\lambda)(1+\mu)} I.$

But $V = (I-i_3)R = \dfrac{IR(2+2\lambda+2\mu+2\lambda\mu-\lambda-\mu-2\lambda\mu)}{2(1+\lambda)(1+\mu)}.$

Therefore $\dfrac{V}{I} = \dfrac{R(2+1+\mu)}{2(1+\lambda)(1+\mu)} = \dfrac{R}{2}\left(\dfrac{1}{1+\lambda}+\dfrac{1}{1+\mu}\right)$

as before.

Example 6. A network, constructed from wire of uniform resistance per unit length, consists of a regular polygon of n sides, and n straight wires joining the centre O of the polygon to its vertices $A_1, \ldots, A_n$. Each side of the polygon is of resistance σ. The point O is earthed, and the vertex A_1 maintained at potential V. Show that the current in the wire $A_r A_{r+1}$ is

$$i_r = 2V\sigma^{-1}\sin\alpha\sinh(n+1-2r)\alpha\operatorname{sech}n\alpha,$$

where α is defined by $\cosh 2\alpha = 1 + \sin(\pi/n)$.

Suppose that the current system is composed of n circulating currents $i_1, i_2, \ldots, i_n$ as shown in Figs. 5.14 (i) and (ii) together with a current I in the branch A_1O. [This current I is the total current flowing through the network and could be considered as a circulating current round the circuit consisting of the external battery, its leads and A_1O.] Then Kirchhoff's first law is satisfied at all the junctions.

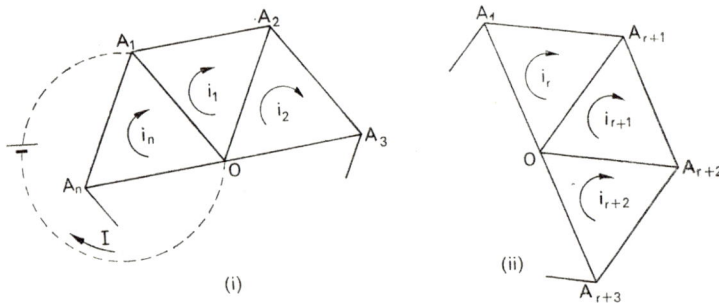

FIG. 5.14

The second law applied to the circuit $OA_{r+1}A_{r+2}Q$ gives

$$(i_{r+1}-i_r)\frac{\sigma}{2}\operatorname{cosec}\left(\frac{\pi}{n}\right)+i_{r+1}\sigma+(i_{r+1}-i_{r+2})\frac{\sigma}{2}\operatorname{cosec}\left(\frac{\pi}{n}\right)=0,$$

i.e.
$$i_{r+2}-2\{1+\sin(\pi/n)\}i_{r+1}+i_r=0, \quad r=1,2,\ldots,n-2. \tag{1}$$

Writing $1+\sin(\pi/n) = \cosh 2\alpha$ the difference equation (1) can be written

$$i_{r+2}-(e^{2\alpha}+e^{-2\alpha})i_{r+1}+i_r=0.$$

The solution of this difference equation can be written

$$i_r = C_1\cosh 2r\alpha + C_2\sinh 2r\alpha, \quad r=1,2,\ldots,n, \tag{2}$$

where C_1, C_2 are constants to be determined.

Clearly, by symmetry, $i_n = -i_1$. Therefore

$$C_1\cosh 2\alpha + C_2\sinh 2\alpha + C_1\cosh 2n\alpha + C_2\sinh 2n\alpha = 0,$$

i.e.
$$C_1\cosh(n+1)\alpha + C_2\sinh(n+1)\alpha = 0. \tag{3}$$

Ohm's law applied to the path A_1A_2O gives

$$i_1\sigma + (i_1-i_2)\frac{\sigma}{2}\operatorname{cosec}\left(\frac{\pi}{n}\right) = V,$$

where V is the potential difference between A and O. Therefore

$$i_1 + (i_1-i_2)\frac{1}{2(\cosh 2\alpha - 1)} = \frac{V}{\sigma},$$

i.e.
$$i_1(2\cosh 2\alpha - 1) - i_2 = 2V(\cosh 2\alpha - 1)/\sigma.$$

Therefore

$$C_1(2\cosh^2 2\alpha - \cosh 2\alpha - \cosh 4\alpha) + C_2(2\cosh 2\alpha\sinh 2\alpha - \sinh 2\alpha - \sinh 4\alpha)$$
$$= 2V(\cosh 2\alpha - 1)/\sigma,$$

§ 5.9 THE STEADY FLOW OF ELECTRIC CURRENTS

i.e.
$$C_1(1-\cosh 2\alpha) - C_2 \sinh 2\alpha = 2V(\cosh 2\alpha - 1)/\sigma$$
or
$$C_1 \sinh \alpha + C_2 \cosh \alpha = -(2V \sinh \alpha)/\sigma. \tag{4}$$

Solution of eqns. (3) and (4) gives
$$\frac{C_1}{\sinh(n+1)\alpha \sinh \alpha} = \frac{C_2}{-\cosh(n+1)\alpha \sinh \alpha} = \frac{2V/\sigma}{\cosh n\alpha}.$$

Then substituting in eqn. (2)
$$i_r = \frac{2V \sinh \alpha}{\sigma \cosh n\alpha} \{\sinh(n+1)\alpha \cosh 2r\alpha - \cosh(n+1)\alpha \sinh 2r\alpha\}$$
$$= 2(V/\sigma) \sinh \alpha \sinh(n+1-2r)\alpha \operatorname{sech} n\alpha.$$

Example 8. A telegraph wire $A_0 A_1 A_2 \ldots A_n$ has a number of faulty connections at $A_1, A_2, \ldots, A_{n-1}$ through which the current leaks to earth. The portion of the wire between A_k and A_{k+1} ($0 \leq k \leq n-1$) has a resistance r whilst the resistance between A_k ($1 \leq k \leq n-1$) and the earth is R. Prove that the potential V_k of the fault A_k is
$$V_k = \frac{V_0 \sinh(n-k)\theta + V_n \sinh k\theta}{\sinh n\theta},$$
where $\cosh \theta = 1 + \dfrac{r}{2R}$.

If the terminal A_n is earthed, and $n\theta$ is large, show that the effective resistance between A_0 and the earth is approximately
$$\tfrac{1}{2}[r + (r^2 + 4rR)^{1/2}].$$

Since we need V_k, we express the equations of the problem in terms of the potentials at $A_0, A_1, \ldots, A_n$ (Fig. 5.15). Ohm's law gives currents flowing as follows:

Element	Current
$A_k A_{k+1}$,	$I_k = (V_k - V_{k+1})/r$;
$A_{k+1} A_{k+2}$,	$I_{k+1} = (V_{k+2} - V_{k+1})/r$;
A_{k+1} to earth,	$J_{k+1} = V_{k+1}/R.$

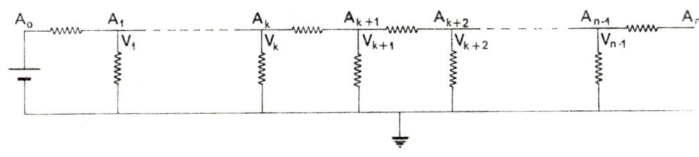

Fig. 5.15

Kirchhoff's first law applied to the joint A_{k+1} gives
$$I_k = I_{k+1} + J_{k+1}, \quad k = 0, 1, \ldots, n-1,$$
from which it follows that
$$V_{k+2} - \left(2 + \frac{r}{R}\right) V_{k+1} + V_k = 0,$$
i.e.
$$V_{k+2} - 2\cosh \theta \, V_{k+1} + V_k = 0.$$

The solution of this difference equation is
$$V_k = A \cosh k\theta + B \sinh k\theta, \quad k = 0, 1, \ldots, n.$$
Fitting the potentials at $k = 0$, $k = n$ gives
$$A = V_0, \quad A \cosh n\theta + B \sinh n\theta = V_n$$
and so, after some reduction,
$$V_k = \frac{V_0 \sinh (n-k)\theta + V_n \sinh k\theta}{\sinh n\theta}.$$
The total current I flowing in the circuit is that in the branch $A_0 A_1$,

i.e.
$$I = \frac{V_0 - V_1}{r}.$$
When $V_n = 0$, we find
$$I = \frac{V_0}{r}\left[1 - \frac{\sinh(n-1)\theta}{\sinh n\theta}\right] = \frac{V_0}{r}\left[1 - \left(\frac{e^{(n-1)\theta} - e^{-(n-1)\theta}}{e^{n\theta} - e^{-n\theta}}\right)\right]$$
$$= \frac{V_0}{r}\left[1 - \frac{e^{-\theta}(1 - e^{-2(n-1)\theta})}{1 - e^{-2n\theta}}\right]$$
$$\approx \frac{V_0}{r}(1 - e^{-\theta}) \text{ when } n\theta \text{ is large.}$$

But $\cosh \theta = 1 + \dfrac{r}{2R}$, $\sinh \theta = \sqrt{(\cosh^2 \theta - 1)} = \sqrt{\left\{\dfrac{r}{2R}\left(2 + \dfrac{r}{2R}\right)\right\}}$.

Therefore $\quad e^{-\theta} = 1 + \dfrac{r}{2R} - \dfrac{1}{2R}\sqrt{(r^2 + 4rR)}$.

Therefore $\quad I \approx \dfrac{V_0}{2Rr}\{\sqrt{(r^2 + 4rR)} - r\}$.

But if S is the effective resistance between A_0 and earth, $I = V_0/S$.

Therefore $\quad S \approx \dfrac{2Rr}{\sqrt{(r^2 + 4rR)} - r} = \dfrac{1}{2}\{r + \sqrt{(r^2 + 4rR)}\}$.

Exercises 5.9

1. Twelve wires each of resistance r form the edges of a cube. Current enters and leaves at the ends of one of the wires. Determine the equivalent resistance of the cube.
2. The sides of a hexagon $ABCDEF$ are formed of wires of resistance r. The points A, C, E are connected to a junction G by wires each of resistance s and the points B, D, F are connected to a junction H by wires each also of resistance s. Prove that the resistance to a current that enters the network at A and leaves at D is
$$\frac{r(r+9s)}{6(r+s)}.$$
3. A generator of direct current producing a voltage V is connected to a radiator of variable resistance R by means of two wires of fixed combined resistance r. Show that in order to obtain the maximum possible output of heat from the radiator R must be chosen to equal r. Sketch the dependence of the heat output on R and find the biggest permissible value of R if at least half the maximum possible output is to be obtained.

4. Coils of resistances R_1 and R_2 ($R_1 \geqslant R_2$) are connected, in parallel, to a battery of internal resistance r. Show that the rate of expenditure of energy in the two coils will exceed that in either of the coils if connected separately to the battery if

$$\frac{1}{r^2} > \frac{1}{R_2}\left(\frac{1}{R_1}+\frac{1}{R_2}\right).$$

5. Terminals A, B, C, D are connected by resistances r_1, r_2, r_3, r_4 in AB, AC, BD, CD respectively, and a galvanometer of negligible resistance is connected across BC. A current I enters at A and leaves at D and the current through the galvanometer (in the direction from B to C) is i. Show that, if $r_3/r_4 = \alpha$ and $r_1/r_2 = \alpha(1-\delta)$, where δ is small, then δ is given approximately by

$$\frac{i}{I}\frac{(\alpha+1)^2}{\alpha}.$$

6. Three terminals A, B, C in a network of linear conductors are connected by wires BC, CA and AB of resistances a, b, c respectively. These wires are then removed and A, B, C are connected to a new common terminal P by wires of resistances α, β, γ respectively. Show that the distribution of currents in the remainder of the network will be unaltered if

$$a\alpha = b\beta = c\gamma = abc/(a+b+c).$$

In a certain network the terminals X, Y, Z, W are connected by wires YZ, ZX, XY, XW, YW, ZW of resistances 1, 2, 3, 1, 2, 3 units respectively. Show that the equivalent resistance between X and Y is $(39/40)$ units.

7. Four resistors A, B, C, D of equal nominal value have a common terminal. The other terminal of resistor D is earthed. If the true value of each resistor is known to lie between $r-\delta$ and $r+\delta$, find the limits between which the potential difference across the resistor D must lie when:
   (i) the free terminals of resistors A, B, C are all at potential V;
   (ii) one free terminal is earthed and the other two are at potential V;
   (iii) two free terminals are earthed and the other one is at potential V.
If δ is small, show that the ranges of variation in the three cases are respectively $\tfrac{3}{4}a$, a, $\tfrac{3}{4}a$, where $a = V\delta/r$.

8. A network consists of six similar uniform conducting wires forming the edges of a regular tetrahedron $PQRS$. A current j enters the network at P and leaves at a point K on QR where $QK/QR = \lambda$. Show that the current in PS is independent of λ and find the currents in the other wires.

9. Terminals A, B, C, D are connected by a galvanometer having an adjustable resistance r in AB and fixed resistances R_1, R_2, R_3 in BC, CD, DA. B and D are also connected by a resistance r', and an external circuit containing a battery of e.m.f. E and a total resistance R is connected across A, C. Prove that, if r is adjusted so that there is no current in r', the current in the galvanometer is

$$ER_2/\{R(R_1+R_2)+R_1(R_2+R_3)\}.$$

10. A generator of direct current producing a constant voltage is connected to an electric radiator by two wires each of resistance r. The radiator consists of two heating elements each of resistance 40 ohms. It is observed that the power output of the radiator when the two elements are in parallel is 49/16 times as great as when they are in series. What is the resistance r of the connecting wires?

11. A network, in the form of a quadrilateral and one diagonal, is made up of wires AB, BC, CD, DA of resistances a, b, c, d respectively, A and C being also joined by a wire AC of resistance e. Currents i, j, k are led into the network at A, B, D respectively, a

current $i+j+k$ leaving the network at C. Prove that the current in the wire AC is

$$\frac{i(a+b)(c+d)+jb(c+d)+kc(a+b)}{e(a+b+c+d)+(a+b)(c+d)}.$$

12. Four terminals A, B, C, D, arranged in a square, are connected by resistances such that P is the resistance between A and B, R between B and C, S between C and D, Q between D and A, U between B and D. A potential difference V is maintained between A and C. Find the current through U.
If $Q = P$ and $S = R+X$, where X is small compared with all other resistances mentioned, show that for fixed R, U, X, V, the value of P leading to maximum current through U is approximately

$$P = R\left[\frac{U}{U+2R}\right]^{1/2}.$$

13. A network of uniform wire has the shape of a rectangle of sides a, na, divided into n square meshes of side a by $n-1$ parallel wires, the current I entering and leaving by adjacent corners of the rectangle, which are also corners of the first mesh. If the cyclic current in the rth mesh is I_r, show that

$$I_{r-1}-4I_r+I_{r+1} = 0, \qquad r = 1, 2, 3, \ldots, n,$$

where $I_0 = I$ and $I_{n+1} = 0$, and hence deduce that the equivalent resistance R of the network is

$$r(1-\alpha)(1+\alpha^{2n+1})/(1-\alpha^{2n+2}),$$

where $\alpha = 2+\sqrt{3}$, and r is the resistance of a length a of the wire.
Show also that $\sqrt{3}-1 < R/r < 3/4$.

14. $A_0A_1A_2 \ldots$ and $B_0B_1B_2 \ldots$ are two semi-infinite parallel wires. Each element A_rA_{r+1} and B_rB_{r+1} has resistance R ohms. Joining each A_rB_r is a wire of resistance R ohms. Current I enters the network at A_0 and flows off to infinity along the wires. If V_r is the potential at A_r and U_r is the potential at B_r, prove that if $r \neq 0$,

$$3V_r - V_{r-1} - V_{r+1} - U_r = 0,$$
$$3U_r - U_{r-1} - U_{r+1} - V_r = 0.$$

If $W_r = V_r - U_r$, show that $W_r = \dfrac{RI}{1+\sqrt{3}}(2-\sqrt{3})^r$.

By considering a source of current at A_0 and a corresponding sink at B_0, prove that the resistance of the network between A_0B_0 is $2R/(1+\sqrt{3})$.

15. Direct current is carried from a generator G to a load of resistance R by a single wire with an earth return. The wire is supported by imperfect insulators at points A_1, A_2, $\ldots$, A_{n-1}. The resistance of each insulator is r, and the resistance between G and A_1 of each wire A_kA_{k+1} and betwen A_{n-1} and the load is ϱ. The currents in GA_1, $A_{k-1}A_k$ and through the load are i_1, i_k, i_n respectively.
Show that

$$i_k \sinh (n-1)\alpha = i_1 \sinh (n-k)\alpha + i_n \sinh (k-1)\alpha,$$
$$k = 1, 2, \ldots, n$$

and that

$i_1\varrho \cosh \tfrac{1}{2}\alpha = i_n[\varrho \cosh (n-\tfrac{1}{2})\alpha + 2R \sinh \tfrac{1}{2}\alpha \sinh (n-1)\alpha]$, where $4 \sinh^2 \tfrac{1}{2}\alpha = \varrho/r$.

16. A uniform wire A_0A_n has resistance nr and is connected to earth by wires each of resistance kr at the $(n-1)$ points A_1, A_2, $\ldots$, A_{n-1}, where $A_0A_1 = A_1A_2 = \ldots = A_{n-1}A_n$.

A_0 is maintained at potential ϕ_0 and A_n is earthed. If ϕ_s is the potential at A_s, show that

$$\phi_{s+1} - 2(\cosh \alpha)\phi_s + \phi_{s-1} = 0,$$

where
$$\cosh \alpha = 1 + \frac{1}{2k}.$$

Express ϕ_s/ϕ_0 as a function of α, and find the current in $A_{n-1}A_n$.

5.10 Integral theorems applied to networks

All the theorems discussed in § 5.8 can be applied to networks.

1. The uniqueness theorem

For a network this theorem states that, given the following:

(i) the resistance of every branch (i.e. the wire joining two nodes) of the network,

(ii) the e.m.f. of the battery in every branch (this may, of course, be zero for some branches),

(iii) either the potential of the node, or the current strength entering the network there, for every node connected to any external sources of current; then the current strength in every branch of the network is uniquely determined.

2. The reciprocal theorem

The form taken by eqn. (5.20) when applied to a network can lead to the following result. We may regard *every* node of the network as a possible electrode at which current may enter the system; if, in one distribution, the

Fig. 5.16

potentials of the nodes are V_r ($r = 1, 2, \ldots$) and the currents entering are I_r, and in a second distribution the corresponding quantities are V'_r and I'_r, then, as in eqn. (5.20),

$$\sum_r V_r I'_r = \sum_r V'_r I_r. \qquad (5.29)$$

We can extend this result to include branches which contain sources of e.m.f. Suppose the branch joining nodes i and j contains an e.m.f. ψ_{ij}; then we regard the terminals of the battery A_i, A_j (see Fig. 5.16) as additional nodes. Current $-I_{ij}$ enters the network at A_i and current I_{ij} enters at A_j, their potentials being ψ_i, ψ_j where

$$\psi_j - \psi_i = \psi_{ij}. \tag{5.30}$$

If in the second distribution the e.m.f. ψ_{ij} is replaced by an e.m.f. ψ'_{ij}, and the corresponding current is I'_{ij}, then we have additional terms for the new nodes occurring in the summations in eqn. (5.29). The left-hand side becomes $\sum_r V_r I'_r + \sum_A (\psi_j - \psi_i) I'_{ij}$, and the right-hand side becomes $\sum_r V'_r I_r + \sum_A (\psi'_j - \psi'_i) I_{ij}$.

Because of the relation (5.30) the amended form of eqn. (5.29) now is

$$\sum_r V_r I'_r + \sum_A \psi_{ij} I'_{ij} = \sum_r V'_r I_r + \sum_A \psi'_{ij} I_{ij}, \tag{5.31}$$

where the sum $\sum_A$ is taken over all the extra nodes introduced, i.e. over all those branches which contain batteries.

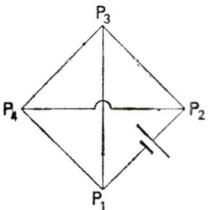

Fig. 5.17

Example. The resistances in the network of Fig. 5.17 are such that when a battery is inserted in $P_1 P_2$ no current flows in $P_3 P_4$. If the battery in $P_1 P_2$ is removed and one is inserted in $P_3 P_4$, then no current flows in $P_1 P_2$.

3. The Heating Effect

We apply the technique used in § 5.9 to obtain Kirchhoff's second law when we consider this result. The general form of the theorem is eqn. (5.21), where the integral is taken through the whole of the conducting volume. For a typical branch AB of the network:

$$\iiint \frac{j^2 \, d\tau}{\sigma} = \int_A^B \frac{i_{AB}^2 \alpha \, ds}{\alpha^2 \sigma} = i_{AB}^2 \int_A^B \frac{ds}{\sigma \alpha} = i_{AB}^2 R_{AB}.$$

Hence the total heating effect in the whole network is $H = \sum i_{AB}^2 R_{AB}$, where the sum is taken over all branches.

4. THE MINIMUM THEOREM

The most general form of this theorem is given in eqn. (5.25). The integral there, which has a minimum, is taken through the whole conducting volume. The first member gives the sum obtained above, $\sum i_{AB}^2 R_{AB}$. For a typical branch AB the second member gives

$$\iiint \boldsymbol{j} \cdot \boldsymbol{\mathcal{E}} \, d\tau = \int_A^B \left(\frac{i_{AB}}{\alpha}\right) \alpha \boldsymbol{\mathcal{E}} \cdot d\boldsymbol{s} = i_{AB} \int_A^B \boldsymbol{\mathcal{E}} \cdot d\boldsymbol{s} = i_{AB}\psi_{AB}.$$

Therefore we conclude that the currents are distributed in the network in such a way that the sum

$$H = \sum (i_{AB}^2 R_{AB} - 2i_{AB}\psi_{AB}) \quad \text{is a minimum.} \tag{5.32}$$

(In reducing the above integrals we have taken the current density to be effectively uniform over the cross-section of the wire, i.e. $|\boldsymbol{j}| = i/\alpha$, and the volume element $d\tau$ has been replaced by αds, where ds is an element of arc along the wire, so that the volume integration is reduced to the single integration along the wire, taken over the lengths of all the branches in the network.)

Example 1. Six conducting wires form a hexagon $ABCDEFA$, and two other wires join B to F, and C to E. Each wire has resistance R. A battery of electromotive force V and negligible resistance is inserted in AB. Show that the heat generated per second is $11V^2/(30R)$.

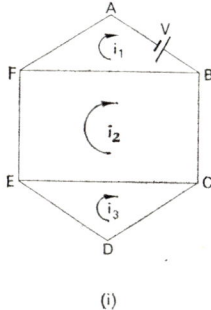 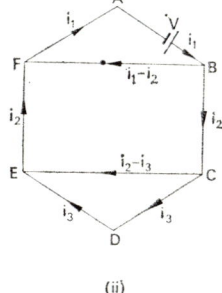

FIG. 5.18

Here we choose circulating currents as shown in Fig. 5.18 (i); the actual currents flowing in the various members of the network are shown in Fig. 5.18 (ii).

Applying Kirchhoff's second law we obtain for the stated circuits:

$ABFA$: $i_1 R + (i_1 - i_2)R + i_1 R = V, \quad \therefore \quad 3i_1 - i_2 = V/R;$ (1)

$BCEFB$: $i_2 R + (i_2 - i_3)R + i_2 R - (i_1 - i_2)R = 0, \quad \therefore \quad i_1 - 4i_2 + i_3 = 0;$ (2)

$CDEC$: $i_3 R + i_3 R - (i_2 - i_3)R = 0, \quad \therefore \quad i_2 = 3i_3.$ (3)

Solution of eqns. (1)–(3) for i_1 gives $i_1 = 11V/(30R)$.

The energy supplied per second to the circuit by the battery is $Vi_1 = 11V^2/(30R)$ and this must be the heat generated per second.

Alternatively the quantity $H = \sum (i_{AB}^2 R_{AB} - 2i_{AB}\psi_{AB})$ in this case becomes
$$H = R\{3i_1^2 + 4i_2^2 + 3i_3^2 - 2i_1 i_2 - 2i_2 i_3 - 2i_1(V/R)\}.$$
We obtain the conditions for a stationary value of H:

$$\frac{1}{R}\frac{\partial H}{\partial i_1} = 6i_1 - 2i_2 - 2\frac{V}{R} = 0, \quad \frac{1}{R}\frac{\partial H}{\partial i_2} = 8i_2 - 2i_1 - 2i_3 = 0,$$

$$\frac{1}{R}\frac{\partial H}{\partial i_3} = 6i_3 - 2i_2 = 0.$$

These equations are equivalent to those obtained above.

Example 2. Prove that in a network of linear conductors in which a given current enters at a point A and leaves at a point B, the currents are distributed in such a way that the loss of energy in heating the conductors is a minimum, subject to the condition that there is no accumulation of charges at any junction of the wires.

If the network consists of n similar wires in parallel, joining A and B, apply the above theorem to show that the current along each of the wires is the same.

The first paragraph of this example is a restatement of the theorem of p. 185 applied to a network of wires.

In the particular case of n similar wires, each of resistance R, connected in parallel, suppose that the currents in the respective wires are $i_1, i_2, \ldots, i_n$ and that the total current flowing is I. Then the theorem implies that

$$H = \sum_{r=1}^{n} Ri_r^2$$

is a minimum subject to the condition

$$\sum i_r = I \text{ (constant)}.$$

Then, using Lagrange's undetermined multipliers,

$$\delta H = 2\sum_{r=1}^{n} Ri_r \delta i_r = 0,$$

$$\sum_{r=1}^{n} \delta i_r = 0.$$

Therefore
$$\sum_{r=1}^{n} (2Ri_r + \lambda)\delta i_r = 0,$$

where λ is the (undetermined) Lagrangian multiplier. It follows that

$$i_r = -\frac{\lambda}{2R},$$

and so the current along each wire is the same I/n.

Example 3. $ABCDA$ is a closed circuit of uniform wire of total resistance $2R$, the lengths ABC, CDA being equal. Another wire joining A and C contains a battery of electromotive force E, and another wire joining B and D contains a galvanometer. The resistance of the battery and the wire connecting it to A and C is R_b and that of the galvanometer and the wire connecting it to B and D is R_g. Show that, if the difference between the resistances of AB and AD is a given small quantity r, whose square may be neglected, the position of B on the wire ABC for which the current through the galvanometer is least is the middle point.

Show further that the least current is given by

$$2Er/\{(2R_b+R)(2R_g+R)\}.$$

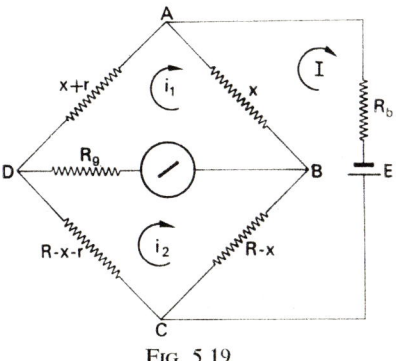

FIG. 5.19

We take the resistance of AB as x. The configuration is then as shown in Fig. 5.19 which also gives the circulating currents. Applying Kirchhoff's second law we obtain for the stated circuits:

$ACBA$: $R_b I+(R-x)(I-i_2)+x(I-i_1) = E$;
$ABDA$: $x(i_1-I)+R_g(i_1-i_2)+(x+r)i_1 = 0$;
$BCDB$: $(R-x)(i_2-I)+(R-x-r)i_2+R_g(i_2-i_1) = 0$.

These equations may be rewritten

$$(R_b+R)I-xi_1-(R-x)i_2 = E, \qquad (1)$$
$$-xI+(2x+R_g)i_1-R_g i_2 = -ri_1, \qquad (2)$$
$$-(R-x)I-R_g i_1+(2R-2x+R_g)i_2 = ri_2. \qquad (3)$$

Equivalent equations can be derived by minimising

$$H' = (x+r)i_1^2+(R-x-r)i_2^2+x(I-i_1)^2+(R-x)(I-i_2)^2+R_g(i_1-i_2)^2+R_b I^2-2EI,$$

considered as a function of i_1, i_2, I. [The appropriate equations are $\partial H'/\partial I = \partial H'/\partial i_1 = \partial H'/\partial i_2 = 0$.]

We solve eqns. (1)–(3) by an iterative process. A first approximation $(I)_0$, $(i_1)_0$, $(i_2)_0$ is obtained by neglecting the terms on the right-hand sides of eqns. (2) and (3). Noting that by symmetry $(i_1)_0 = (i_2)_0 = \frac{1}{2}(I)_0$ we find

$$(I)_0 = 2(i_1)_0 = 2(i_2)_0 = \frac{E}{(\frac{1}{2}R+R_b)}.$$

The next approximation $(I)_1$, $(i_1)_1$, $(i_2)_1$ is found by putting $i_1 = i_2 = \frac{1}{2}I_0 = E/(R+2R_b)$ on the right-hand sides of eqns. (2) and (3) and solving the resulting equations (2a) and (3a) together with eqn. (1). Taking eqns. 2(1)+(2a)+(3a) gives $(I)_1 = (I)_0$ and it follows from eqns. (1) and (2a) that

$$x(i_1)_1+(R-x)(i_2)_1 = \frac{ER}{R+2R_b}, \qquad (4)$$

$$(2x+R_g)(i_1)_1-R_g(i_2)_1 = \frac{(2x-r)E}{R+2R_b}. \qquad (5)$$

Then $2x(4) - R(5)$ gives

$$i_1 - i_2 = \frac{RrE}{(2x^2 - 2xR - RR_g)(R + 2R_b)}.$$

It follows that the current $i_1 - i_2$ through the galvanometer is least when

$$|2x^2 - 2xR - 2RR_g| = \left| 2\left(x - \frac{R}{2}\right)^2 - \frac{R^2}{2} - RR_g \right|$$

is greatest, i.e. when $x = R/2$ and B is the mid-point of AC. In this case

$$|i_1 - i_2| = 2Er/\{(2R_g + R)(2R_g + R)\}.$$

[This arrangement of four resistors $ABCD$ together with a galvanometer and a battery is known as a Wheatstone's bridge. The investigation given in the above problem estimates the sensitivity of the arrangement for detecting the "balance" when the current through the galvanometer is zero.]

Miscellaneous Exercises V

1. Six uniform wires, each of resistance R, are connected at their ends so that they form the edges of a tetrahedron $ABCD$. Points E, F on BC, AD respectively are such that $CE = \lambda CB$ and $AF = \mu AD$. Show that the equivalent resistance of the network between E and F is

$$\tfrac{1}{2} R\{1 + \lambda + \mu - \lambda^2 - \mu^2\}.$$

2. Four points P_1, P_2, P_3, P_4 are joined by six conducting wires 12, 13, 14, 23, 24, 34, where 12, for instance, denotes the wire joining P_1 and P_2; the resistances of the wires are R_{12}, R_{13}, ... and there are no batteries in the network. If a current C enters at P_1 and leaves at P_2 and current C' enters at P_3 and leaves at P_4, show that the condition that there shall be no current in 24 is

$$\begin{vmatrix} R_{12}, & R_{14}, & CR_{12} \\ R_{23}, & R_{34}, & C'R_{34} \\ R_{13} + R_{23}, & -(R_{13} + R_{14}), & 0 \end{vmatrix} = 0.$$

3. A, B, C are three branch points of a network of linear conductors. The resistances of the branches BC, CA, AB are respectively a, b, c. These branches are removed and the points A, B, C are connected to a common centre O by resistances α, β, γ respectively. Show that, if

$$a\alpha = b\beta = c\gamma = \beta\gamma + \gamma\alpha + \alpha\beta,$$

then there will be no change in the distribution of current in the remainder of the network.

Points $A_1, \ldots, A_n$ are connected in pairs by wires each of resistance R. By generalizing the previous result, or otherwise, show that the equivalent resistance between any pair of points is $2R/n$.

4. A telegraph wire connects the terminals A_0 and A_n and makes contact with terminals $A_1, A_2, \ldots, A_{n-1}$ in order. The portion of the wire between A_k and A_{k+1} ($0 \leq k \leq n-1$) has a resistance r, while between A_k ($1 \leq k \leq n-1$) and earth there is a resistance R.

§ 5.10   THE STEADY FLOW OF ELECTRIC CURRENTS   189

Prove that the potential V_k of the terminal A_k is given by

$$V_k = \frac{V_0 \sinh(n-k)\theta + V_n \sinh k\theta}{\sinh n\theta},$$

where $\cosh\theta = 1 + r/2R$.
If the terminal A_n is insulated from earth (so that $V_{n-1} = V_n$) find the value of V_n in terms of V_0. Show that in this case, if $n\theta$ is large, the effective resistance between A_0 and earth is

$$r/(1 - e^{-\theta}).$$

5. A certain resisting element which does not obey Ohm's law has the property that the magnitude of the current which flows through it is k times the fourth power of the voltage applied across it. This unit is connected in series with a resistance R across the terminals of a direct current generator which develops a voltage V. Show that if $R = 8/kV^3$ the voltage v across the element is equal to $\tfrac{1}{2}V$ and that if V now changes by a small amount δV (R and k remaining fixed), v changes by only $\tfrac{1}{8}\delta V$.

6. A uniform telephone wire AB of total resistance R is connected to earth by means of a resistance r at an unknown point X of its length such that $AX = x \cdot AB$. It is found by measurement that the effective resistance between the end A and earth is $\lambda_1 R$ when B is insulated, and is $\lambda_2 R$ when B is earthed. Show that $1 - \lambda_2$ and $\lambda_1 - \lambda_2$ are both necessarily positive, and prove that the position of X is determined by the quadratic equation

$$x^2 - 2\lambda_2 x + \lambda_1\lambda_2 - \lambda_1 + \lambda_2 = 0.$$

Show that the appropriate root is the smaller of the two roots of this equation.

7. A network of equivalent resistance r_1 is connected in parallel with a resistance r_2. Find the equivalent resistance of the system.
A uniform wire $A_0 A_n$ is divided by the points A_i ($i = 1, \ldots, n-1$) into n equal parts, each of resistance r. A parallel identical wire $B_0 B_n$ is similarly divided. A ladder-shaped network is formed by joining A_i and B_i ($i = 1, \ldots, m$) by equal wires of resistance R. Current enters at A_0 and leaves at B_0, and the equivalent resistance of the network is R_m. Prove that

$$R_{m+1} = 2r + \frac{RR_m}{R + R_m}.$$

Show also that, if m is large, the equivalent resistance of the network is approximately $r + \sqrt{(r^2 + 2rR)}$.

8. An infinite two-dimensional square grid of wires consists of elements, each of resistance R ohms, i.e. a wire of R ohms resistance connects the point (m, n) to each of the points $(m-1, n)$, $(m+1, n)$, $(m, n-1)$, $(m, n+1)$ for all integral values of m, n. A current I enters the grid at the point $(0,0)$. Prove that the potential $V_{m,n}$ of the grid-point (m, n) satisfies the difference equation

$$4V_{m,n} - V_{m-1,n} - V_{m,n-1} - V_{m,n+1} = RI\delta_{m0}\delta_{n0},$$

where $\delta_{m0} = 0$ for $m \neq 0$ and $\delta_{00} = 1$.
Verify that a solution of this equation is

$$V_{m,n} = \frac{RI}{8\pi^2} \int_{-\pi}^{\pi}\int_{-\pi}^{\pi} \frac{e^{imx+iny} - 1}{2 - \cos x - \cos y}\,dx\,dy.$$

By considering a source of current at $(0, 0)$ and a sink of current at $(1, 0)$, prove that the resistance between two adjacent grid-points is $\tfrac{1}{2}R$.

9. A current flows between two electrodes. Show that the rate of generation of heat is less for the natural current distribution than any other distribution consistent with the conservation law and the same total current.

    A perfectly conducting sphere is contained within another and the space between is filled with a homogeneous conducting medium. Prove that the resistance between the spheres is a maximum when they are concentric.

10. The resistance of a cable is R per unit length and the conductance of the insulation is $1/S$ per unit length. Show that, in the steady state, the potential V and the current i at a distance x from one end of the cable satisfy the relations

    $$V = -S\frac{di}{dx}, \qquad -\frac{dV}{dx} = Ri.$$

    If the end $x = 0$ is kept at a constant potential E and the other end $x = l$ is earthed, and if $R = r(1+\varepsilon x/l)$, $S = s(1+\varepsilon x/l)$, where r and s are constant and ε is small, show that

    $$V = E(1+\tfrac{1}{2}\varepsilon x/l)\operatorname{cosech} nl \sinh n(l-x)$$

    approximately, where $n = (r/s)^{1/2}$.

11. Deduce, from the force acting on a moving charge, that the rate of heat production per unit volume at a point r inside a distribution of conducting material is $j^2(r)/\sigma(r)$, where $j(r)$ and $\sigma(r)$ represent the current density and the conductivity at r.

    A steady current $I\,(\neq 0)$ enters a system of bodies of finite conductivity (not necessarily wires) at an electrode A and leaves at an electrode B. Show that $j(r)/I$ is a vector independent of I.

    The potential of the electrode at A is V_A and the potential of the electrode at B is V_B. Show that the total rate of heat production throughout the system of conducting bodies is $I(V_A - V_B)$. Show also that $(V_A - V_B)/I$ is independent of I.

    A cavity is excavated within the conducting material. Show that, if I is maintained constant, the total rate of heat production does not decrease, whereas, if $V_A - V_B$ is maintained constant, the total rate of heat production does not increase.

ANSWERS TO THE EXERCISES

Exercises 2.7 (p. 50)

2. They form the conjugate system of (non-intersecting) coaxal circles, all being orthogonal to the lines of force.

Exercises 2.12 (p. 69)

1. $\mathbf{M}\cdot\mathbf{r}/(4\pi\varepsilon_0|\mathbf{r}|^2)$, where $\mathbf{r}$ is the position vector of a field point referred to the dipole as origin. Since the field is two-dimensional, $\mathbf{r}$ is a two-dimensional vector; using plane polar coordinates with the dipole at the origin, and pointing along $\theta = 0$, the equipotential surfaces are the circular cylinders $r = \lambda_1 \cos\theta$, the lines of force lie in the (orthogonal) circular cylinders $r = \lambda_2 \sin\theta$.

3. $\frac{1}{3}P$.

Exercises 3.3 (p. 89)

1. $4\pi\varepsilon_0 ab/(b-a)$; $\quad 4\pi\varepsilon_0 a$; $\quad 4\pi\varepsilon_0 \left\{\dfrac{ab}{b-a} + \dfrac{bc}{c-b}\right\}$.

2. Outermost $V_3 = \dfrac{e_1+e_2+e_3}{4\pi\varepsilon_0 c}$; next $V_2 = V_3 + \dfrac{e_1+e_2}{4\pi\varepsilon_0}\left(\dfrac{1}{b} - \dfrac{1}{c}\right)$;

innermost $V_1 = V_2 + \dfrac{e_1}{4\pi\varepsilon_0}\left(\dfrac{1}{a} - \dfrac{1}{b}\right)$.

3. $\dfrac{(e_1+e_2)}{4\pi\varepsilon_0 b} + \dfrac{e_1}{4\pi\varepsilon_0}\left(\dfrac{1}{r} - \dfrac{1}{b}\right)$, where r is distance from the centre of the spheres.

Miscellaneous Exercises III (p. 109)

2. *Modified theorem.* If a system of charges e_i placed at points P_i gives potentials V_j to a system of uncharged conductors, then charges e'_j placed on the conductors raise the points P_i to potentials V'_i, where $\sum e_i V'_i = \sum e'_j V_j$; $e/(4\pi\varepsilon_0 c)$.

5. The increase in energy of the condenser is equal to the energy supplied by the battery.

6. See p. 100; $2\pi\varepsilon_0 v^2 (c-b) a^2 b / \{c(b-a)^2\}$.

7. $3Z^2 e^2/(20\pi\varepsilon_0 R)$.

8. $\dfrac{Q_1 Q_2 k^2 c}{8\pi^2 \varepsilon_0 (ab)^{3/2}} \displaystyle\int_0^{\pi/2} \dfrac{d\theta}{(1-k^2 \sin^2\theta)^{3/2}}$.

10. $\varepsilon_0 V^2 A/(2d^2)$.

Exercises 4.2 (p. 119)

1. $8\pi\varepsilon_0 a$.

2. $\dfrac{\varepsilon_0(S-W/d')}{d} + \dfrac{\varepsilon_0 W}{d'\left\{d-d'\left(1-\dfrac{1}{K}\right)\right\}}$.

3. $4\pi\varepsilon_0 \bigg/ \left(\dfrac{1}{Ka} - \dfrac{1}{Kc} + \dfrac{1}{c} - \dfrac{1}{b}\right)$.

6. $2\pi\varepsilon_0 \bigg/ \left\{\dfrac{1}{K_2}\ln b - \dfrac{1}{K_1}\ln a + \left(\dfrac{1}{K_1} - \dfrac{1}{K_2}\right) \sum_{s=1}^{2n-1}(-1)^{s-1} \ln\left(a + \dfrac{s(b-a)}{2n}\right)\right\}$.

Exercises 4.3 (p. 129)

4. $-Q \bigg/ \left\{4\pi r^2 K_1 \varepsilon_0 \left(1 + \dfrac{1}{K_2} + \dfrac{2}{K_1}\right)\right\}$ for $a < r < 2a$;

   $-Q \bigg/ \left\{4\pi r^2 K_2 \varepsilon_0 \left(1 + \dfrac{1}{K_2} + \dfrac{2}{K_1}\right)\right\}$ for $2a < r < 4a$;

   $Q\left(\dfrac{1}{K_2} + \dfrac{2}{K_1}\right) \bigg/ \left\{4\pi r^2 \varepsilon_0 \left(1 + \dfrac{1}{K_2} + \dfrac{2}{K_1}\right)\right\}$ for $r > 4a$.

5. Zero on inner surface of sphere $r = a$, $q_1/(4\pi a^2)$ on outer surface of sphere $r = a$, $-q_1/(4\pi b^2)$ on inner surface of sphere $r = b$, $(q_1+q_2)/(4\pi b^2)$ on outer surface of sphere $r = b$;

$$V = \dfrac{q_1}{4\pi\varepsilon_0}\left(\dfrac{1}{r} + \dfrac{\lambda r}{b^2}\right) + \dfrac{(q_2 - \lambda q_1)}{4\pi\varepsilon_0 b}.$$

Miscellaneous Exercises IV (p. 146)

1. $2\pi\varepsilon_0 \left\{\dfrac{1}{K}\ln\left(\dfrac{a+d}{a}\right) + \ln\left(\dfrac{c}{a+d}\right)\right\}$.

2. $V = a^3\varrho/(3\varepsilon_0 r)$, $E_r = a^3\varrho/(3\varepsilon_0 r^2)$ for $r > a$;

   $V = \dfrac{\varrho a^2}{6\varepsilon_0}\left(2 + \dfrac{1}{K}\right) - \dfrac{\varrho r^2}{6K\varepsilon_0}$, $E_r = \varrho r/(3K\varepsilon_0)$.

3. (ii) C where $\dfrac{1}{C} = \dfrac{1}{K_{12}C_{12}} + \dfrac{1}{K_{23}C_{23}} + \ldots + \dfrac{1}{K_{n-1,n}C_{n-1,n}}$.

Exercises 5.7 (p. 164)

2. $2/(3\lambda a^2)$.

4. $V = (bV_a - aV_b)/(b-a) - (V_a - V_b)r/(b-a)$.

6. $\dfrac{1}{\pi\sigma t}\ln\left(\dfrac{b}{a}\right)$.

7. (i) $\sqrt{(Rr)}\tanh nl$;
   (ii) $\sqrt{(Rr)}\coth nl$.

Exercises 5.9 (p. 180)

1. $7r/12$.   3. $(3+\sqrt{8})r$.
7. (i) $\dfrac{3V}{2}\left(\dfrac{r\pm\delta}{2r\pm\delta}\right)$;   (ii) $\dfrac{V}{2}\left(1\pm\dfrac{\delta}{r}\right)$;   (iii) $\dfrac{V}{2}\left(\dfrac{r\pm\delta}{2r\pm\delta}\right)$.
8. PQ, $\tfrac{1}{4}j(2-\lambda)$; PR, $\tfrac{1}{4}j(1+\lambda)$; PS, $\tfrac{1}{4}j$;
   SQ, $\tfrac{1}{4}j(1-\lambda)$; SR, $\tfrac{1}{4}j\lambda$; RK, $\tfrac{1}{4}j(1+2\lambda)$;
   QK, $\tfrac{1}{4}j(3-2\lambda)$.
12. $V(RQ-PS)/\{PR(Q+S)+(P+R)(QS+QU+US)\}$.
16. $\phi_s/\phi_0 = [\sinh\{(n-s)\alpha\}]/\sinh(n\alpha)$; $(\phi_0 \sinh\alpha)/\{r \sinh(n\alpha)\}$.

Miscellaneous Exercises V (p. 188)

4. $V_n = (V_0 \sinh\theta)/\{\sinh n\theta - \sinh(n-1)\theta\}$.
7. $r_1 r_2/(r_1+r_2)$.

INDEX

Alternating current   8
Ampere   5, 8
Anisotropic medium   115
Atom   11, 12
Avogadro's number   5

Capacitance   4, 29, 79, 85, 111
Capacitor   3, 81, 84
CAVENDISH   30
Cavity, field inside   65, 67
Champion's experiment   12
Cloud-chamber   14
Coefficient
  of capacitance   79
  of influence   79
Condenser   3, 76, 81, 84, 111
Conductivity   152, 154
Conductors   2, 11, 14, 71, 79, 111, 149, 152
Conservation
  of charge   21, 150, 151, 154
  of energy   25
Corkscrew   7
Coulomb, the   23
Coulomb's law   6, 29, 30, 31, 37, 40, 43
Current   4, 12, 150
Current density   150

Dielectric   3, 14, 146
Dielectric constant   111, 116, 122
Dielectric susceptibility   116
Dipole   5, 7, 58, 64, 92, 112
Displacement current   22
Displacement, electric   26, 21
Double layer   61

Electrode   149, 151, 153, 154, 156, 159, 166

Electrolysis   4, 5, 156
Electromagnetic induction   8
Electromagnetic oscillation   10
Electromotive force   156, 157
Electron   5, 11, 12, 14, 152
  conduction   11
Electrophorus   73
Electroscope   2
Energy   90, 94, 101, 103, 105, 130, 137
Equipotential surface   26, 71, 72

Faraday   3, 8, 23, 80, 111, 113, 122
Faraday's law of electrolysis   5
Faraday's law of electromagnetic induction   8, 21
Field line   25
Field-strength, electric   24, 31
FIZEAU   17
Fleming's rule   8
FRESNEL   16

Galilean transformation   14
Galvanic cell   5, 156
Galvanometer   14
Gauss's theorem   40, 47, 53, 55, 58, 71, 117, 118, 119, 132
Green's reciprocal theorem   83, 120
Green's theorem   55, 75, 83, 121, 166, 167

Harmonic function   74
Heat production (by a current)   155, 168, 170, 184
Hertz   8, 10, 15, 22

"Ice-pail" experiment   3, 23, 80
Induction   3, 80
Induction coil   9

INDEX

Influence 3, 28, 80, 82
Insulator 2, 11
Interference 10
Isotropic medium 115

KELVIN 121, 137, 166, 167
Kelvin's minimum energy theorem 137, 169
Kirchhoff's laws 172, 173, 184

Laplace's equation 39, 73, 74, 83, 160
Laws of motion 1, 16
Leyden jar 3
Line of force 7, 25, 80
Line-density of charge 24, 33, 36
Linear (medium) 115, 121, 132, 138

Magnetism 5
Magnetometer 5, 6
MAXWELL 9, 10, 15, 22, 144
Maxwell's stress tensor 144
Mole 5
Moment of dipole 58
Multipole 58

Network of conductors 172

Oersted 7
Oersted's experiment 7
Ohm 5, 151
Ohm's law 4, 151, 152, 153, 154, 158

Permittivity 29, 115
Poisson's equation 41, 54, 55, 57, 73
Poisson's (equivalent) distribution 66, 67, 68, 112
Polarization 10, 15, 64, 112, 113, 114, 120, 138, 155, 158, 169
Poles 6
Ponderomotive forces 91, 100, 111, 143

Potential 4, 25, 29, 31, 56, 73, 75, 79, 104, 153
Principal stress 145

Quadrupole 60

Reciprocal theorem 83, 120, 167, 183
Relative permittivity 116
Resistance 4, 151, 159
RIEMANN 9

SI (Système International) 1, 6
Screening (electrostatic) 80, 84
Solenoid 7
Specific inductive capacity 111
Stream-lines 150, 161
Stress 141, 143
Superposition, principle of 31, 68
Surface curl 54
Surface density of charge 24
Surface divergence 53

Test charge 24
Transformer 8
Tube
  of flow 150
  of force 46, 141, 142, 146, 150

Uniqueness theorem 74, 75, 120, 166, 183
Unit tubes 47

Valency 5
Velocity of light 16
Voltage 4
Volume density of charge 24

WILSON AND WILSON 14